# Studies in Computational Intelligence

## Volume 591

**Series editor**

Janusz Kacprzyk, Polish Academy of Sciences, Warsaw, Poland
e-mail: kacprzyk@ibspan.waw.pl

**About this Series**

The series "Studies in Computational Intelligence" (SCI) publishes new developments and advances in the various areas of computational intelligence—quickly and with a high quality. The intent is to cover the theory, applications, and design methods of computational intelligence, as embedded in the fields of engineering, computer science, physics and life sciences, as well as the methodologies behind them. The series contains monographs, lecture notes and edited volumes in computational intelligence spanning the areas of neural networks, connectionist systems, genetic algorithms, evolutionary computation, artificial intelligence, cellular automata, self-organizing systems, soft computing, fuzzy systems, and hybrid intelligent systems. Of particular value to both the contributors and the readership are the short publication timeframe and the world-wide distribution, which enable both wide and rapid dissemination of research output.

More information about this series at http://www.springer.com/series/7092

Jia Zeng · Zhi-Qiang Liu

# Type-2 Fuzzy Graphical Models for Pattern Recognition

Jia Zeng
School of Computer Science
  and Technology
Soochow University
Suzhou
China

Zhi-Qiang Liu
School of Creative Media
City University of Hong Kong
Hong Kong
China

ISSN 1860-949X
ISBN 978-3-662-51522-8
DOI 10.1007/978-3-662-44690-4

ISSN 1860-9503 (electronic)
ISBN 978-3-662-44690-4 (eBook)

Springer Heidelberg New York Dordrecht London

Jointly published with Tsinghua University Press, Beijing
ISBN: 978-7-302-36890-8, Tsinghua University Press, Beijing

Printed on acid-free paper

Springer is part of Springer Science+Business Media (www.springer.com)

*This book is dedicated to my family: Tian-Yi Zeng and Xiao-Qin Cao*

# Preface

This state-of-the-art book describes important advances in type-2 fuzzy systems that have been made in the past decade for real-world pattern recognition problems, such as speech recognition, handwriting recognition, and topic modeling. The success of type-2 fuzzy sets has been largely attributed to their three-dimensional membership functions to handle both randomness and fuzziness uncertainties in real-world problems. In pattern recognition, both features and models have uncertainties, such as nonstationary babble noise in speech signals, large variations of handwritten Chinese character shapes, uncertain meaning of words in topic modeling, and uncertain parameters of models because of insufficient and noisy training data. All these uncertainties motivate us to integrate type-2 fuzzy sets with probabilistic graphical models to achieve better overall performance in terms of robustness, generalization ability, or recognition accuracy. For example, we integrate type-2 fuzzy sets with graphical models such as Gaussian mixture models, hidden Markov models, Markov random fields, and latent Dirichlet allocation-based topic models for pattern recognition. The type-2 fuzzy Gaussian mixture models can describe uncertain densities of observations. The type-2 fuzzy hidden Markov models incorporate the first-order Markov chain into the type-2 fuzzy Gaussian mixture models, which is suitable for modeling uncertain speech signals under babble noise. The type-2 fuzzy Markov random fields combine type-2 fuzzy sets with Markov random fields, which is able to handle large variations in structural patterns such as handwritten Chinese characters. The type-2 fuzzy topic models focus on uncertain mixed membership of words to different topical clusters, which is effective to partition the observed (visual) words into semantically meaningful topical themes. In conclusion, these real-world pattern recognition applications demonstrate the effectiveness of type-2 fuzzy graphical models for handling uncertainties.

Suzhou, May 2013
Hong Kong

Jia Zeng
Zhi-Qiang Liu

# Contents

1   **Introduction**. . . . . . . . . . . . . . . . . . . . . . . . . . . . . . . . . . . . . . . . .   1
    1.1   Pattern Recognition . . . . . . . . . . . . . . . . . . . . . . . . . . . . . .   1
    1.2   Uncertainties . . . . . . . . . . . . . . . . . . . . . . . . . . . . . . . . . . .   3
    1.3   Book Overview . . . . . . . . . . . . . . . . . . . . . . . . . . . . . . . . . .   6
    References . . . . . . . . . . . . . . . . . . . . . . . . . . . . . . . . . . . . . . . . . . .   7

2   **Probabilistic Graphical Models** . . . . . . . . . . . . . . . . . . . . . . . . . . .   9
    2.1   The Labeling Problem . . . . . . . . . . . . . . . . . . . . . . . . . . . . .   9
    2.2   Markov Properties . . . . . . . . . . . . . . . . . . . . . . . . . . . . . . . . 10
    2.3   The Bayesian Decision Theory . . . . . . . . . . . . . . . . . . . . . . . 11
        2.3.1   Descriptive and Generative Models . . . . . . . . . . . . . . . 14
        2.3.2   Statistical–Structural Pattern Recognition . . . . . . . . . . . 15
    2.4   Summary . . . . . . . . . . . . . . . . . . . . . . . . . . . . . . . . . . . . . . 15
    References . . . . . . . . . . . . . . . . . . . . . . . . . . . . . . . . . . . . . . . . . . . 16

3   **Type-2 Fuzzy Sets for Pattern Recognition** . . . . . . . . . . . . . . . . . . 17
    3.1   Type-2 Fuzzy Sets. . . . . . . . . . . . . . . . . . . . . . . . . . . . . . . . 17
    3.2   Operations on Type-2 Fuzzy Sets. . . . . . . . . . . . . . . . . . . . . . 22
    3.3   Type-2 Fuzzy Logic Systems . . . . . . . . . . . . . . . . . . . . . . . . 25
        3.3.1   Fuzzifier. . . . . . . . . . . . . . . . . . . . . . . . . . . . . . . . . . 25
        3.3.2   Rule Base and Inference. . . . . . . . . . . . . . . . . . . . . . . 27
        3.3.3   Type Reducer and Defuzzifier. . . . . . . . . . . . . . . . . . . 28
    3.4   Pattern Recognition Using Type-2 Fuzzy Sets . . . . . . . . . . . . . 32
    3.5   The Type-2 Fuzzy Bayesian Decision Theory . . . . . . . . . . . . . 35
    3.6   Summary . . . . . . . . . . . . . . . . . . . . . . . . . . . . . . . . . . . . . . 39
    References . . . . . . . . . . . . . . . . . . . . . . . . . . . . . . . . . . . . . . . . . . . 42

**4 Type-2 Fuzzy Gaussian Mixture Models** .................... 45
    4.1 Gaussian Mixture Models ............................. 45
    4.2 Type-2 Fuzzy Gaussian Mixture Models................. 48
    4.3 Multi-category Pattern Classification .................. 53
    References ................................................ 55

**5 Type-2 Fuzzy Hidden Moarkov Models**..................... 57
    5.1 Hidden Markov Models ............................. 57
        5.1.1 The Forward-Backward Algorithm................. 59
        5.1.2 The Viterbi Algorithm ......................... 60
        5.1.3 The Baum–Welch Algorithm ..................... 61
    5.2 Type-2 Fuzzy Hidden Markov Models ................... 62
        5.2.1 Elements of a Type-2 FHMM...................... 64
        5.2.2 The Type-2 Fuzzy Forward-Backward Algorithm ...... 64
        5.2.3 The Type-2 Fuzzy Viterbi Algorithm............... 66
        5.2.4 The Learning Algorithm......................... 67
        5.2.5 Type-Reduction and Defuzzification .............. 71
        5.2.6 Computational Complexity ....................... 71
    5.3 Speech Recognition ................................. 72
        5.3.1 Automatic Speech Recognition System ............. 72
        5.3.2 Phoneme Classification ......................... 74
        5.3.3 Phoneme Recognition .......................... 78
    5.4 Summary ......................................... 79
    References ................................................ 82

**6 Type-2 Fuzzy Markov Random Fields**...................... 85
    6.1 Markov Random Fields ............................. 85
        6.1.1 The Neighborhood System ....................... 87
        6.1.2 Clique Potentials.............................. 88
        6.1.3 Relaxation Labeling............................ 89
    6.2 Type-2 Fuzzy Markov Random Fields ................... 91
        6.2.1 The Type-2 Fuzzy Relaxation Labeling ............. 94
        6.2.2 Computational Complexity ....................... 96
    6.3 Stroke Segmentation of Chinese Character ................ 96
        6.3.1 Gabor Filters-Based Cyclic Observations ........... 97
        6.3.2 Stroke Segmentation Using MRFs................. 100
        6.3.3 Stroke Extraction of Handprinted Chinese Characters.... 102
        6.3.4 Stroke Extraction of Cursive Chinese Characters....... 102
    6.4 Handwritten Chinese Character Recognition ............... 105
        6.4.1 MRFs for Character Structure Modeling............. 109
        6.4.2 Handwritten Chinese Character Recognition (HCCR).... 115
        6.4.3 Experimental Results........................... 119
    6.5 Summary ......................................... 124
    References ................................................ 127

**7  Type-2 Fuzzy Topic Models** ............................ 129
    7.1  Latent Dirichlet Allocation ......................... 129
        7.1.1  Factor Graph for the Collapsed LDA .............. 132
        7.1.2  Loopy Belief Propagation (BP) .................. 134
        7.1.3  An Alternative View of BP ..................... 137
        7.1.4  Simplified BP (siBP) .......................... 139
        7.1.5  Relationship to Previous Algorithms ............ 139
        7.1.6  Belief Propagation for ATM .................... 141
        7.1.7  Belief Propagation for RTM .................... 143
    7.2  Speedup Topic Modeling ............................ 146
        7.2.1  Fast Topic Modeling Techniques ................ 147
        7.2.2  Residual Belief Propagation ................... 148
        7.2.3  Active Belief Propagation ..................... 149
    7.3  Type-2 Fuzzy Latent Dirichlet Allocation ............. 155
        7.3.1  Topic Models ................................. 158
        7.3.2  Type-2 Fuzzy Topic Models (T2 FTMs) ........... 161
    7.4  Topic Modeling Performance ........................ 168
        7.4.1  Belief Propagation ............................ 168
        7.4.2  Residual Belief Propagation ................... 175
        7.4.3  Active Belief Propagation ..................... 179
    7.5  Human Action Recognition .......................... 189
        7.5.1  Feature Extraction and Vocabulary Formation ..... 190
        7.5.2  Results on KTH Data Set ...................... 191
    References ............................................ 195

**8  Conclusions and Future Work** ......................... 199
    8.1  Conclusions ....................................... 199
    8.2  Future Works ...................................... 201

**Errata to: Type-2 Fuzzy Graphical Models for Pattern
        Recognition** ..................................... E1

# Acronyms

| | |
|---|---|
| ABP | Active belief propagation |
| BP | Belief propagation |
| GLM | Generalized linear model |
| GMM | Gaussian mixture model |
| HMM | Hidden Markov model |
| LDA | Latent Dirichlet allocation |
| L-LDA | Labeled latent Dirichlet allocation |
| MF | Membership functions |
| MRF | Markov random field |
| PDF | Probabilistic density function |
| RBP | Residual belief propagation |
| T2 FGMM | Type-2 fuzzy Gaussian mixture model |
| T2 FHMM | Type-2 fuzzy hidden Markov model |
| T2 FMRF | Type-2 fuzzy Markov random field |
| T2 FS | Type-2 fuzzy sets |

# Chapter 1
# Introduction

**Abstract** This chapter overviews the whole book. First, we introduce some fundamental concepts in pattern recognition. Pattern recognition can be viewed as a labeling process that bridges human (machine) perceptions to linguistic labels. Second, we motivate the use of probabilistic graphical models and type-2 fuzzy sets to handle two important uncertainties, namely randomness and fuzziness, existing universally in the labeling problem. Finally, we summarize our contributions, and provide the structure of this book.

## 1.1 Pattern Recognition

The term pattern recognition encompasses a wide range of information processing problems of great practical significance, from speech recognition to the classification of handwritten characters [1]. By patterns, we understand any relations, regularities, or structure inherent in some source of data [7]. Pattern recognition takes in raw data and makes decisions based on the "category" of the pattern [3], which deals with the automatic detection of patterns in data, and plays a crucial role in many modern artificial intelligence and computer science problems.

As shown in Fig. 1.1, the pattern recognition system includes five basic components: (1) sensing, (2) segmentation, (3) feature extraction, (4) classification, and (5) post-processing, where the components (1)–(3) simulate the human perception leading to the *feature space*, and the components (4)–(5) assign the linguistic labels to features for classification. From Fig. 1.1, we may view pattern recognition as a labeling process that uses *linguistic labels* (classes) to interpret the machine perception. Mathematically, the pattern recognition system reflects a functional relationship between the input and the output decision. This function is sometimes referred to as the *decision function*. Usually, we will choose a particular set or class of candidate functions known as *hypotheses* before we begin trying to determine the correct function. The ability of a hypothesis to correctly classify data not in the training set is known as its *generalization*. The process of determining the correct function (often

© Tsinghua University Press, Beijing and Springer-Verlag Berlin Heidelberg 2015
J. Zeng and Z.-Q. Liu, *Type-2 Fuzzy Graphical Models for Pattern Recognition*,
Studies in Computational Intelligence 591, DOI 10.1007/978-3-662-44690-4_1

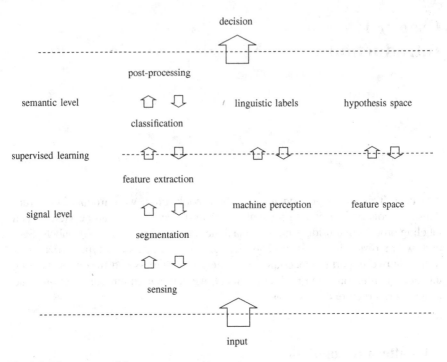

decision

post-processing

semantic level          linguistic labels          hypothesis space

classification

supervised learning

feature extraction

signal level          machine perception          feature space

segmentation

sensing

input

**Fig. 1.1**  The structure of the pattern recognition system

a number of adjustable parameters) on the basis of examples of input/output func-
tionality is called *learning* or *training*. When the examples are input/output pairs,
it is called *supervised learning*. The examples for training are generally referred to
as the *training data*. The *learning algorithm* takes the training data as input, and
selects a hypothesis from the *hypothesis space*. So the learning algorithm connects
the feature space and the hypothesis space. Figure 1.1 shows two layers, signal level
and semantic level, of the pattern recognition system. They correspond to machine
perception and linguistic labels as well as feature space and hypothesis space.

Based on the above, pattern recognition involves three central problems:

1. How to extract features so that the feature space can be partitioned efficiently;
2. How to choose the set of hypotheses so that the hypothesis space contains the
   correct representation of the decision function;
3. How to design the learning algorithm to automatically determine the decision
   function from the feature space and hypothesis space.

To solve above problems, we have to incorporate knowledge about the problem
domain called *prior knowledge*. The choice of the distinguishing features is to achieve
a "good" *pattern representation*, and depends on the characteristics of the problem
domain. The representation may naturally reveal the structural relationships among
the components, and express the true underlying model of the patterns. We favor

a small number of features, which may lead to both simpler decision regions and learning algorithm for classifiers [3]. The second problem is associated with the choice of functions that can best describe the variations of features within class. In this book, we assume a parametric form of the decision function so that the hypothesis space has a specific functional form with undetermined parameters. Hence prior knowledge about distinguishing features and the functional form of hypothesis plays a major role in successful pattern recognition systems. Therefore, the learning algorithm in the context of this book determines the parameters of the decision function based on the training data. The learning process adapts the hypothesis to fit the training data. In the meanwhile, we hope that the hypothesis is not only consistent with the training data, but also generalizes well to the unknown test data.

This book mainly focuses on handling uncertainties in pattern recognition. Specifically, we combine two techniques, probabilistic graphical models [4] and type-2 fuzzy sets [5], to handle some sources of uncertainties, and hope to improve the performance of pattern recognition systems.

## 1.2 Uncertainties

Inevitably, pattern recognition has to deal with uncertainties. In statistical pattern recognition, we focus on the statistical properties of patterns in terms of *randomness*, which is generally expressed in probability density functions (PDFs). The features are often called *observations*. The success of statistical pattern recognition has been largely due to its ability to recover the model that generated the patterns. It assumes the models which give rise to the data do not themselves evolve with time, i.e., the sources of data are stationary. Hence the probability densities with parameters estimated from a large amount of training data are enough to represent the random uncertainty of patterns. Note that the sufficient training data play key roles to characterize randomness in both practical and theoretical aspects.

Graphical models [2, 4, 6] use Markov properties as special hypotheses that can statistically represent the structural relationships in observations. They define the most important structural information by the neighborhood system, and encode such information in terms of randomness in the PDFs. To achieve the lowest probability of classification error, the Bayesian decision theory [3] provides the optimal decision boundary as shown in Fig. 1.2a, b. Two graphical models, namely hidden Markov models (HMMs) and Markov random fields (MRFs), have been widely explored as hypotheses for modeling sequential and two-dimensional patterns, respectively. The difference between HMMs and MRFs lies in their neighborhood systems. For example, HMMs are suitable acoustic models for phonemes, because HMMs reflect phonemes piece-wise stationary properties. On the other hand, MRFs are good at describing the stroke relationships of handwritten characters. The main advantages of using graphical models are twofold: Markov properties are able to model patterns statistical-structurally in terms of randomness. Besides, graphical models have efficient leaning and decoding algorithms with a tractable computational complexity.

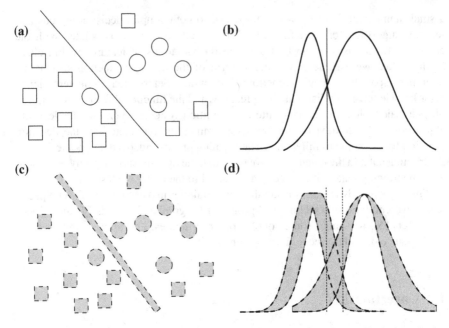

**Fig. 1.2**  The uncertainties exist in both the feature and hypothesis spaces. In **a** and **b**, the hypothesis space describes the randomness in the feature space, and the decision boundary is a *solid line* for two pattern classes, namely *circles* and *squares*. In **c** and **d**, after incorporating fuzziness, the fuzzy (*the shaded region*) hypothesis space describes both randomness and fuzziness in the fuzzy feature space, and the decision boundary becomes fuzzy denoted by the *shaded line*

However, in practice, we often encounter uncertainties that cannot be characterized by randomness as follows.

1. Uncertain Feature Space: Features (training and test data) may be corrupted by noise. Measurement noise is non-stationary, and the mathematical description of the non-stationarity is unknown (e.g., as in a time-varying signal-to-noise ratio (SNR)) (Fig. 1.2c);
2. Uncertain Hypothesis Space: Insufficient or noisy training data may result in uncertain parameters of the hypothesis, so that the decision boundary is also uncertain (Fig. 1.2d);
3. Non-stationarity: Features have statistical attributes that are non-stationary, and the mathematical descriptions of the non-stationarity are unknown.

All of these uncertainties can be considered as *fuzziness* resulting from incomplete prior knowledge, i.e., fuzzy features (we do not know if the training data are clean, stationary, and sufficient), fuzzy hypothesis (we do not know if the parameters imply the correct mapping), and fuzzy attributes (we do not know the attributes exactly).

Type-2 fuzzy sets is a good choice to describe the *bounded uncertainty* in both the feature and hypothesis spaces. Type-2 fuzzy sets have grades of membership

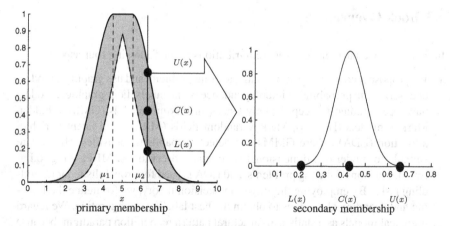

**Fig. 1.3** A type-2 membership grade can be any subset in [0, 1]—the primary membership and corresponding to each primary membership, there is a secondary membership (which can also be in [0, 1]) that defines the possibilities for the primary membership. The *left* shows the uncertainty of the primary membership. The *right* shows the corresponding secondary membership of one of the bounded vertical-slice primary memberships $[L(x), U(x)]$ given the input $x$

that are themselves fuzzy. A type-2 membership grade can be any subset in [0, 1]—the primary membership and corresponding to each primary membership, there is a secondary membership (which can be in [0, 1]) that defines the possibilities for the primary membership [5]. Figure 1.3 shows an example of the type-2 fuzzy set. The primary membership is uncertain denoted by the bounded shaded region on the left, where its uncertainty is measured by the secondary membership on the right. Note that at each input $x$, the uncertainty of primary membership is reflected by the bounded interval $[L(x), U(x)]$. Hence, type-2 fuzzy sets provide a natural framework to simultaneously handle more uncertainties thanks to their three-dimensional membership functions. Using primary memberships, we may handle the randomness. On the other hand, we can handle the fuzziness of primary memberships by secondary memberships. Both randomness and fuzziness can propagate in the system through type-2 fuzzy set operations. By incorporating type-2 fuzzy sets, we can describe the fuzzy feature and hypothesis in terms of bounded uncertainty. We have two major advantages of using type-2 fuzzy sets: they can characterize patterns in terms of both randomness and fuzziness; meanwhile, type-2 fuzzy sets retain a controlled degree (bounded) of uncertainty based on prior knowledge.

In conclusion, on the one hand, we motivate the use of graphical models to represent structural patterns statistically. On the other hand, we incorporate type-2 fuzzy sets to handle both randomness and fuzziness within a unified framework.

## 1.3 Book Overview

In this book, we have made several contributions in the following four aspects:

1. We propose pattern recognition as the labeling problem, and use graphical models to describe the probabilistic interdependence of the labels. Four graphical models have been studied in deep, namely Gaussian mixture models (GMMs), hidden Markov models (HMMs), Markov random fields (MRFs), and latent Dirichlet allocation (LDA), where GMMs can detect probabilistic densities, HMMs are suitable to model one-dimensional sequential observations, MRFs are good at two-dimensional labeling problems, and LDA is used for probabilistic topic modeling tasks. By employing the Bayesian decision theory, we formulate the learning and decoding algorithms to obtain the best labeling configuration. We regard graphical models as a statistical-structural pattern recognition paradigm, because they statistically describe the structural information of labels and observations.
2. We investigate the mechanism of type-2 fuzzy sets for handling uncertainties. The three-dimensional membership function enables type-2 fuzzy sets to handle more uncertainties within a unified framework. Two important properties, secondary membership function and foot print of uncertainty, determine the capability of type-2 fuzzy sets for modeling bounded uncertainty. Also we review the recent advances of type-2 fuzzy sets for pattern recognition.
3. The major contribution of this book is that we integrate graphical models with type-2 fuzzy sets referred to as the type-2 fuzzy graphical models to handle both random and fuzzy uncertainties within a unified framework. We have developed the learning and decoding algorithms of the type-2 fuzzy graphical models based on type-2 fuzzy set operations. We show that the type-2 fuzzy graphical models can be viewed as embedded with many classical graphical models, so that the expressive power of type-2 fuzzy graphical models have been greatly enhanced.
4. We extensively explore many pattern recognition applications, such as density estimation, speech recognition, handwritten Chinese character recognition, and topic modeling of visual words for human action recognition. The experimental results are encouraging, which confirm the validity of the proposed type-2 fuzzy graphical models.

The first chapter is an introduction to the principal concepts of pattern recognition, and motivates the use of graphical models and type-2 fuzzy sets to handle uncertainties in pattern recognition. Chapter 2 deals with the labeling problem by using graphical models to model the probabilistic interdependency of labels. Type-2 fuzzy sets and fuzzy logic systems are introduced in Chap. 3. We handle random and fuzzy uncertainties within a unified type-2 fuzzy graphical model framework, and extends the graphical models' learning and decoding algorithms using type-2 fuzzy sets operations. Chapter 4 integrates type-2 fuzzy sets with Gaussian mixture models to estimate densities from noisy and insufficient data. Chapter 5 shows how to use type-2 fuzzy hidden Markov models to handle babble noise in speech recognition. Chapter 6 introduces how to use type-2 fuzzy Markov random fields to differentiate handwritten Chinese characters having similar shapes. Chapter 7 discusses type-2 fuzzy latent Dirichlet allocation and its application to human action recognition. Chapter 8 draws conclusions and envisions future works.

# References

1. Bishop, C.M.: Neural Networks for Pattern Recognition. Oxford University Press, New York (1995)
2. Bishop, C.M.: Pattern Recognition and Machine Learning. Springer, Heidelberg (2006)
3. Duda, R.O., Hart, P.E., Stork, D.G.: Pattern Classification, 2nd edn. Wiley, New York (2001)
4. Koller, D., Friedman, N.: Probabilistic Graphical Models: Principles and Techniques. MIT Press, Massachusetts (2009)
5. Mendel, J.M.: Uncertain Rule-based Fuzzy Logic Systems: Introduction and New Directions. Prentice-Hall, Upper Saddle River (2001)
6. Murphy, K.P.: Machine Learning: A Probabilistic Perspective. MIT Press, Massachusetts (2012)
7. Shawe-Taylor, J., Cristianini, N.: Kernel Methods for Pattern Analysis. Cambridge University Press, Cambridge (2004)

# Chapter 2
# Probabilistic Graphical Models

**Abstract** This chapter introduces probabilistic graphical models as a statistical–structural pattern recognition paradigm. Many pattern recognition problems can be posed as labeling problems to which the solution is a set of linguistic labels assigned to extracted features from speech signals, image pixels, and image regions. Graphical models use Markov properties to measure a local probability on the labels within the neighborhood system. The Bayesian decision theory guarantees the best labeling configuration according to the *maximum a posteriori* criterion.

## 2.1 The Labeling Problem

As mentioned in Chap. 1, many pattern recognition problems can be posed as labeling problems to which the solution is a set of linguistic labels assigned to extracted features from speech signals, image pixels, and image regions. For instance, in speech recognition, we may have labels representing phonemes, and such a label set for the word "cat" would have labels for $/k/$, $/a/$, and $/t/$; in stroke segmentation of Chinese characters, we may have labels representing directions, and each character pixel may be associated with one of labels for horizontal, left-diagonal, vertical, and right-diagonal strokes; in Chinese character recognition, we may have labels representing stroke segments, which constitute different character structures; in topic modeling, we may have labels representing topics, which are the basic thematic components for a document. The labeling problem is shown in Fig. 2.1.

We specify a labeling problem in terms of a set of sites, $1 \le i \le I$, and a set of linguistic labels, $1 \le j \le J$; the $j$th label at any site $i$ is denoted by $s_i = j$. A particular labeling configuration for the whole sites is denoted by $\mathscr{S} = \{s_1, s_2, \ldots, s_I\}$. Note that the system can have the same label $j$ at different sites, and not every label needs to be assigned to sites.

The sites may be successive times, image pixels, or image regions. We call them "regular" sites if they have the natural ordering, as for instance successive times form a one-dimensional sequence, in which sites $i - 1$ and $i + 1$ are before and behind site $i$; image pixels form a lattice, where sites $(i', i - 1)$ and $(i', i + 1)$ are on the left and right side of site $(i', i)$. On the other hand, "irregular" sites have no natural

© Tsinghua University Press, Beijing and Springer-Verlag Berlin Heidelberg 2015
J. Zeng and Z.-Q. Liu, *Type-2 Fuzzy Graphical Models for Pattern Recognition*,
Studies in Computational Intelligence 591, DOI 10.1007/978-3-662-44690-4_2

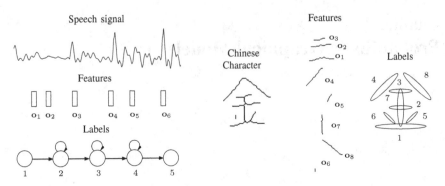

**Fig. 2.1** Many pattern recognition problems can be posed as labeling problems

ordering such as image regions. We can define the ordering of irregular sites when necessary.

The linguistic labels can reflect any relations, regularities, or structures inherent in sites. For instance, phonemes can be divided into three stationary parts— initial, central, and final parts— we may use three labels to represent three successive parts of phoneme data. Chinese characters can be decomposed into blob-level regions (stroke segments), and these regions may be associated with linguistic labels representing character structures. Although labels can take continuous numerical values at each site such as image pixel intensities, we mainly focus on discrete linguistic labels in this book.

## 2.2 Markov Properties

Graphical models assume the label $s_i$ is a random variable at site $i$, thus the labeling configuration $\mathscr{S}$ is a stochastic process. Following the labeling problem, we put a probability measure, namely Markov properties, on the set of all possible labeling configurations,

$$P(\mathscr{S}) > 0, \ \forall \mathscr{S}, \tag{2.1}$$

$$P(s_i|\mathscr{S}_{\{/i\}}) = P(s_i|\mathscr{N}_i), \tag{2.2}$$

where $\mathscr{S}_{\{/i\}}$ are labels at all other sites except $i$, and $\mathscr{N}_i$ are all labels at neighboring sites of $i$. This means that the probability of the label $s_i$ is conditionally independent of all other labels except its neighboring labels. The neighborhood system $\mathscr{N}$ plays an important role to reduce the global measure $P(s_i|\mathscr{S}_{\{/i\}})$ to the local measure $P(s_i|\mathscr{N}_i)$, which may significantly simplify the computational complexity in practice. If we define the neighborhood system at two-dimensional sites, such as image pixels or regions, we call this graphical model the Markov random field (MRF) [1].

If we define the neighborhood system at one-dimensional sites, such as successive times, satisfying $\mathcal{N}_i = i - 1$, that is,

$$P(s_i|s_1, \ldots, s_{i-1}, s_{i+1}, \ldots, s_I) = P(s_i|s_{i-1}), \tag{2.3}$$

we call this graphical model the first-order discrete-time Markov model.

Graphical models represent dependencies among random variables by a graph. Nodes in the graph are equivalent to sites, and edges between nodes imply some relationships between them. Obviously, the MRF is an undirected graphical model with Markov properties, and the first-order discrete-time Markov model has Markov properties on a directed chain graph. Markov properties simplify the global constraints $P(s_i|\mathscr{S}_{\{/i\}})$ for all sites to the local constraints $P(s_i|\mathcal{N}_i)$ for neighboring sites, which greatly reduces the search space to find the best labeling configuration. Moreover, such a simplification is reasonable for piece-wise stationary data in terms of time and space.

Causality is an important property of the neighborhood system $\mathcal{N}$. If $\mathcal{N}_i$ is defined by site $i$'s *previous* sites, it is strictly causal: The probabilities depend only on previous sites. Because regular sites have a natural ordering, it is easy to define the concept "previous." For instance, the $\mathcal{N}$ of the first-order discrete Markov model is causal, because time $i - 1$ always happens before time $i$; the MRF may also have the causal neighborhood system: If we scan the image pixels from left to right and from up to down, then the previous sites of $(i', i)$ are $(i' - 1, i)$ and $(i', i - 1)$. On the other hand, we often define a non-causal neighborhood system at irregular sites, such as image regions, because the concept "previous" at irregular sites is inconsistent in different conditions.

The causal neighborhood systems reduces the search space from "previous" and "following" to only "previous". Such "cause-and-effect" relations are amenable to dynamic programming leading to high computational efficiency. For instance, we can search the best labeling configuration, $\mathscr{S}^* = \{s_1^*, \ldots, s_I^*\}$, for a time sequence as follows: (1) First find the best label $s_1^*$, and then (2) find the best label $s_2^*$ based on the best label $s_1^*$. We repeat this process until $s_I^*$.

## 2.3 The Bayesian Decision Theory

In practice, we have to bridge the gap between labels and data so as to characterize any relations, regularities, or structures inherent in some source of data. After assigning a label $s_i$ to site $i$, we assume that $s_i$ simultaneously generates an observation $\mathbf{o}_i$, which may be symbols, feature vectors, or image pixel values. At each labeling configuration, we will have an observation set, $\mathbf{O} = \{\mathbf{o}_1, \ldots, \mathbf{o}_I\}$, at all sites, where $\mathbf{O}$ belongs to the observation space $\Omega^I$.

In the labeling problem, the task of pattern recognition is equivalent to finding a model $\lambda$ that can provide the best labeling configuration, $\mathscr{S}^* = \{s_1^*, \ldots, s_I^*\}$, to interpret observations, $\mathbf{O} = \{\mathbf{o}_1, \ldots, \mathbf{o}_I\}$. The Bayesian decision theory is a fundamental statistical approach to the problem of pattern recognition. Given a set of observations $\mathbf{O}$ and class models $\lambda_\omega$, we classify $\mathbf{O}$ to the class $\omega^*$ by maximizing the posterior probability,

$$\omega^* = \arg\max_\omega \{P(\omega|\mathbf{O}, \lambda_\omega)\}. \tag{2.4}$$

By employing the Bayes formula,

$$P(\omega|\mathbf{O}, \lambda_\omega) = \frac{P(\mathbf{O}|\lambda_\omega)P(\omega)}{P(\mathbf{O})}, \tag{2.5}$$

where $P(\mathbf{O}|\lambda_\omega)$ is the likelihood of $\lambda_\omega$ given $\mathbf{O}$, and $P(\mathbf{O})$ is a constant normalization factor for all classes $\omega$. For simplicity, we often assume equal prior class probability $P(\omega)$, so that the posterior probability is proportional to the likelihood,

$$P(\omega|\mathbf{O}, \lambda_\omega) \propto P(\mathbf{O}|\lambda_\omega). \tag{2.6}$$

Then Eq. (2.4) becomes

$$\omega^* = \arg\max_\omega P(\mathbf{O}|\lambda_\omega), \tag{2.7}$$

which is the *maximum-likelihood* (ML) criterion.

In graphical models, the class model $\lambda_\omega$ is a set of parameters $\lambda$ and labels $\mathscr{S}$. We assign a labeling configuration $\mathscr{S}$ to the observations $\mathbf{O}$ with a joint probability $P(\mathscr{S}|\lambda, \mathbf{O})$. Thus, the $P(\mathbf{O}|\lambda)$ in Eq. (2.7) is computed by summing over all possible configurations,

$$P(\mathbf{O}|\lambda) = \sum_{\mathscr{S}} P(\mathscr{S}, \mathbf{O}|\lambda). \tag{2.8}$$

Because the direct computation of Eq. (2.8) is usually an intractable combinatorial problem, we can approximate the likelihood $P(\mathbf{O}|\lambda)$ by the most likely labeling configuration, $\mathscr{S}^* = \{s_1^*, \ldots, s_I^*\}$, i.e.,

$$P(\mathbf{O}|\lambda) \approx P(\mathscr{S}^*, \mathbf{O}|\lambda). \tag{2.9}$$

Again the optimal labeling configuration $\mathscr{S}^*$ for the observations $\mathbf{O}$ can be obtained by maximizing the following posterior probability,

$$\mathscr{S}^* = \arg\max_{\mathscr{S}} P(\mathscr{S}|\mathbf{O}, \lambda). \tag{2.10}$$

From the Bayes formula,

$$P(\mathscr{S}|\mathbf{O}, \lambda) = \frac{P(\mathscr{S}, \mathbf{O}|\lambda)}{P(\mathbf{O}|\lambda)} = \frac{p(\mathbf{O}|\mathscr{S}, \lambda)P(\mathscr{S}|\lambda)}{P(\mathbf{O}|\lambda)}, \qquad (2.11)$$

where $P(\mathbf{O}|\lambda)$ is a constant normalization factor for all configurations $\mathscr{S}$. Thus, we obtain

$$P(\mathscr{S}|\mathbf{O}, \lambda) \propto p(\mathbf{O}|\mathscr{S}, \lambda)P(\mathscr{S}|\lambda), \qquad (2.12)$$

where $p(\mathbf{O}|\mathscr{S}, \lambda)$ is the likelihood function for $\mathscr{S}$ given $\mathbf{O}$, and $P(\mathscr{S}|\lambda)$ is the prior probability of $\mathscr{S}$. Therefore, Eq. (2.10) can be rewritten as

$$\mathscr{S}^* = \arg\max_{\mathscr{S}} p(\mathbf{O}|\mathscr{S}, \lambda)P(\mathscr{S}|\lambda). \qquad (2.13)$$

When we have the knowledge about the data distribution but no appreciable prior information about the pattern, we may use the ML criterion to estimate the best labeling configuration. When the situation is the opposite, that is, when we have only prior information, we may use the principle of maximum entropy to find the least biased model that encodes the prior information. With both sources of information available, the best labeling configuration we can get is based on the Bayesian decision theory.

The Bayesian decision theory can be interpreted as a weighting mechanism that weighs the likelihood and prior distributions, and combines them to form the posterior. If these two distributions overlap significantly, this mathematical combination produces a desirable result. Otherwise, it may be possible that the posterior will fall into the region unsupported by either the likelihood or the prior. Therefore, in real applications, we have to balance the likelihood and prior to achieve a desirable labeling configuration.

Note that the likelihood function $p(\mathbf{O}|\mathscr{S}, \lambda)$ is not a probability but a subjective function, which enables us to assign relative weights to different configurations $\mathscr{S}$ given $\mathbf{O}$. On the other hand, the prior probability $P(\mathscr{S}|\lambda)$ is the source of information that exists *prior* to test data in the form of expert judgement and other historical information (training data). For instance, we may specify some structural information in the labeling space based on the knowledge of the problem domain, or just obtain this information from training data automatically. The Bayesian decision rule provides a convenient method to combine two different sources of information, i.e., the data and the prior. It is advantageous to combine distribution functions from different information sources in that the uncertainty in the posterior distribution is reduced when the information in the likelihood and prior distributions are consistent with each other. In other words, the integral information from the data and the prior distribution may have less uncertainty because the data and the prior are from two different information sources that may support each other. From the regularization point of view, the combination of the likelihood and the prior may convert a mathematically ill-posed recognition problem into a well-posed one.

## 2.3.1 Descriptive and Generative Models

We further investigate the graphical models within Bayesian framework in the view of descriptive and generative models.

Descriptive methods construct the model for a pattern by imposing statistical constraints on features extracted from patterns. Linguistic labels can be viewed as high-level features extracted from patterns. Therefore, graphical models are descriptive models, as for instance the first-order discrete Markov model imposes statistical constraints for labels at successive times; the MRF imposes local statistical constraints of labeling configuration at neighboring image pixels or regions. Descriptive models specify the structural constraints $P(s_i|\mathscr{S}_{\{/i\}})$ in the labeling space either specified by prior knowledge or learned through training samples. We may view descriptive constraints as the necessary condition for the pattern class, in which all samples in the pattern class must satisfy these constraints (with high probability), while samples from other classes may also satisfy such constraints. Figure 2.2 shows the relationship between descriptive models and samples, where the circle of descriptive models contains all samples as the necessary condition. Obviously, such a necessary condition is not enough to model the difference among pattern classes, because different pattern classes may probably share the same structure or substructure in the labeling space.

In contrast to descriptive models, generative models are able to randomly generate observed data, typically given some hidden variables at sites. Linguistic labels are a kind of hidden variables, and after randomly assigned to all sites, they simultaneously generate observations **O**. The hidden variables employed to generate observations usually follow very simple models, such as Gaussian mixture models (GMMs). Furthermore, existing generative models appear to suffer from an oversimplified assumption that the observations are independent and identical distributed (i.i.d.)

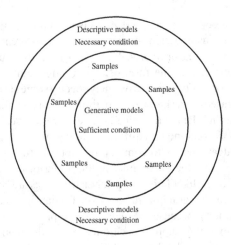

**Fig. 2.2** Descriptive models are the necessary condition and generative models are the sufficient condition for samples from a specific pattern class

for all sites. So the joint probability of observations can be written as a product of probabilities of individual observations,

$$p(\mathbf{O}|\mathscr{S}) = \prod_{i=1}^{I} p(\mathbf{o}_i|s_i = j), \ 1 \leq j \leq J. \tag{2.14}$$

As a result, generative models are not sufficient enough to model realistic patterns. We may view generative models as the sufficient condition for the pattern class, in which the model generates all samples belonging to the pattern class (with high probability), whereas some samples from the pattern class may not be generated by the model (with low probability). Figure 2.2 shows the relationship between generative models and samples, where generative models are subsets of samples as the sufficient condition.

Bayesian decision rule integrates both descriptive and generative models to provide the sufficient and necessary conditions for the pattern class in a hierarchical system. The bottom level of the system is generative in nature, because the observations are generated by hidden variables such as labels. The top level of the system is descriptive in nature, because it governs the relationships among random hidden variables probabilistically. For instance, in speech recognition, we use labels (GMMs) to generate speech feature vectors at each time $i$, and use transition probability $a_{jj'}$ to control the jump from label $j$ to $j'$; in Chinese character recognition, we use labels (GMMs) to generate stroke segments, and use prior clique potentials to encourage or penalize different local labeling configurations.

### 2.3.2 Statistical–Structural Pattern Recognition

The integration of descriptive and generative models also combines both structural and statistical information of the pattern. Descriptive models $P(\mathscr{S})$ can describe the high-level structure of linguistic labels, while generative models $p(\mathbf{O}|\mathscr{S})$ can describe low-level statistical uncertainty of observations. Therefore, graphical models with the Bayesian decision rule provide a theoretically well-founded framework to represent both structural and statistical information existing universally in pattern recognition problems. We call this paradigm the statistical–structural pattern recognition.

## 2.4 Summary

Labeling problems have been proposed to solve problems of computer vision and image analysis in [2]. In this chapter, we extend the same concept to more general problems of pattern recognition. The Bayesian theory has long been the dominant classification methods in pattern recognition, because it rests on an axiomatic foundation that is guaranteed to have quantitative coherence; some other classification methods may not [3]. Further study of the role of the Bayesian theory in fuzzy logic can be found in [4]. The formulation of Bayesian framework for labeling problems is actually the compound Bayes decision problem by the use of context information [3], in which the states of nature are equivalent to the labels here. Markov properties simplify the interdependence of labels by assuming that the labels are only dependent on their neighbors, which avoids the computation of $P(\mathscr{S})$ for all $J^I$ possible values of labeling configuration. Therefore, graphical models have been widely applied to problems of pattern recognition [3], such as speech recognition, handwriting recognition, gesture recognition, face recognition, human motion recognition, and DNA sequence recognition.

The relationship between descriptive models and generative models has been discussed in [5, 6]. We use this idea to justify the modeling ability of graphical models for labeling problems. Graphical models can integrate both descriptive and generative models so that they satisfy the sufficient and necessary conditions to model samples from pattern classes. Murphy and Smyth considered Markov models are special cases of graphical models [7–9], in which HMMs and MRFs are acyclic directed graphs and undirected graphs with Markov properties, respectively. The graph represents the structure of random variables, so graphical models can represent structural patterns statistically [10]. Taking higher order of statistical dependencies between labels into account, graphical models can indirectly reflect statistical dependencies between observations, despite the conditionally independent assumption upon observations in terms of the labels.

# References

1. Li, S.Z., Jain, A.K. (eds.): Handbook of Face Recognition. Springer, New York (2005)
2. Li, S.Z.: Markov Random Field Modeling in Image Analysis. Springer-Verlag, New York (2001)
3. Duda, R.O., Hart, P.E., Stork, D.G.: Pattern Classification, 2nd edn. Wiley, New York (2001)
4. Ross, T.J., Booker, J.M., Parkinson, W.J.: Fuzzy Logic and Probability Applications: Bridging the Gap, 2nd edn. Society for Industrial and Applied Mathematics, Philadelphia (2002)
5. Zhu, S.C.: Statistical modeling and conceptualization of visual patterns. IEEE Trans. Pattern Anal. Machine Intell. **25**(6), 691–712 (2003)
6. Guo, C.E., Zhu, S.C., Wu, Y.N.: Modeling visual patterns by integrating descriptive and generative methods. International Journal of Computer Vision **53**(1), 5–29 (2003)
7. Murphy, K.: A brief introduction to graphical models and Bayesian networks (1998). http://www.cs.ubc.ca/ murphyk/Bayes/bnintro.html
8. Smyth, P.: Belief networks, hidden Markov models, and Markov random fields: a unifying view. Pattern Recognition Letters **18**(11–13), 1261–1268 (1997)
9. Lauritzen, S.L.: Graphical Models. Clarendon Press, Oxford (1996)
10. Zeng, J., Liu, Z.Q.: Markov random fields-based statistical character structure modeling for Chinese character recognition. IEEE Trans. Pattern Anal. Machine Intell. **30**(5), 767–780 (2008)

# Chapter 3
# Type-2 Fuzzy Sets for Pattern Recognition

**Abstract** This chapter covers type-2 fuzzy sets (T2 FSs) and fuzzy logic systems (FLSs). We focus on the mechanism of T2 FSs for handling uncertainties in pattern recognition. Uncertainties exist in both *model* and *data*. Due to the three-dimensional structure of type-2 membership functions (T2 MFs), we may use one dimension, the primary MF, as the hypothesis function, and use another dimension, the secondary MF, to evaluate the uncertainty of the primary MF. At the same time, we may use the type-2 nonsingleton fuzzification to describe the data uncertainty by T2 FSs. We introduce the type-2 fuzzy Bayesian decision theory to handle fuzziness and randomness uncertainties in pattern recognition.

## 3.1 Type-2 Fuzzy Sets

Figure 3.1 shows all terminology of a T2 FS. The T2 FS, denoted $\tilde{A}$, is characterized by a T2 MF $h_{\tilde{A}}(x, u)$, where $x \in X$ and $u \in J_x \subseteq [0, 1]$. Two important concepts distinguish the T2 MF from T1 MF: *secondary MF* and *footprint of uncertainty* (FOU). The secondary MF is a *vertical slice* of $h_{\tilde{A}}(x, u)$ at each value of $x = x'$, i.e., the function $h_{\tilde{A}}(x', u)$, $u \in J_x$. The *amplitude* of the secondary MF is called the *secondary grade*. The *domain* $J_x \subseteq [0, 1]$ of the secondary MF is called the *primary membership* of $x$, and $u$ is the *primary grade*. The FOU is a bounded uncertain region in the primary memberships of a T2 FS $\tilde{A}$, and is the union of all primary memberships. An *upper* MF and a *lower* MF are two T1 MFs that are bounds for the FOU denoted by $\bar{h}_{\tilde{A}}(x)$ and $\underline{h}_{\tilde{A}}(x)$, $\forall x \in X$. If $h_{\tilde{A}}(x, u) = 1, \forall x \in X, \forall u \in J_x \subseteq [0, 1]$, the secondary MFs are interval sets, which reflect a uniform uncertainty at the primary memberships of $x$. This special case is called the interval type-2 fuzzy set (IT2 FS). Because all the secondary grades are unity, the IT2 FS is simplified by the interval of upper and lower MFs, i.e., $h_{\tilde{A}}(x) = [\underline{h}_{\tilde{A}}(x), \bar{h}_{\tilde{A}}(x)]$. In the general T2 FS, the membership grade $h_{\tilde{A}}(x)$ of $x \in X$ $\tilde{A}$ is a T1 FS in $[0, 1]$ (See Fig. 3.1), which we refer to as a secondary set. Similarly, in the IT2 FS, the secondary set $h_{\tilde{A}}(x)$ is an IT1 FS.

Based on the definitions above, the T2 FS has vertical-slice (decomposition) and wavy-slice (canonical) representations respectively. In the vertical-slice manner, we

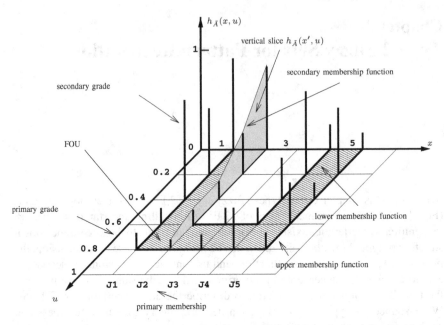

**Fig. 3.1** A T2 MF is three-dimensional. All concepts are labeled here: *vertical* slice, secondary MF, secondary grade, primary MF, primary grade, and lower and upper MFs. The shaded area in the $x-u$ plane is the footprint of uncertainty (FOU)

re-express $\tilde{A}$ by the union of all secondary sets along $x$-axis,

$$\tilde{A} = \{(x, h_{\tilde{A}}(x)) \mid \forall x \in X\}, \tag{3.1}$$

or

$$\tilde{A} = \int_{x \in X} h_{\tilde{A}}(x)/x = \int_{x \in X} \left[ \int_{u \in J_x} f_x(u)/u \right] \Big/ x, \quad J_x \subseteq [0, 1]. \tag{3.2}$$

The FOU is the union of all primary memberships,

$$\text{FOU}(\tilde{A}) = \bigcup_{x \in X} J_x. \tag{3.3}$$

For discrete universes of discourse $X$ and $U$, an *embedded T2 set* $\tilde{A}_e$ has $N$ elements, where $\tilde{A}_e$ contains exactly one element from $J_{x_1}, J_{x_2}, \ldots, J_{x_N}$, namely $u_1, u_2, \ldots, u_N$, each with its associated secondary grade, namely $f_{x_1}(u_1)$, $f_{x_2}(u_2), \ldots, f_{x_N}(u_N)$, i.e.,

$$\tilde{A}_e = \sum_{i=1}^{N'} [f_{x_i}(u_i)/u_i]/x_i, \quad u_i \in J_{x_i} \subseteq U = [0, 1]. \tag{3.4}$$

Set $\tilde{A}_e$ is embedded in $\tilde{A}$, and, there are a total of $\prod_{i=1}^{N} M_i \ \tilde{A}_e$. For discrete universe of discourse $X$ and $U$, an *embedded T1 set* $\tilde{A}_e$ has $N$ elements, one each from $J_{x_1}, J_{x_2}, \ldots, J_{x_N}$, namely $u_1, u_2, \ldots, u_N$, i.e.,

$$A_e = \sum_{i=1}^{N} u_i/x_i \quad u_i \in J_{x_i} \subseteq U = [0, 1]. \tag{3.5}$$

Set $A_e$ is the union of all the primary memberships of set $\tilde{A}_e$ in Eq. (3.4), and there are a total of $\prod_{i=1}^{N} M_i \ A_e$.

In the wavy-slice manner, let $\tilde{A}_e^j$ denote the $j$th T2 embedded set for T2 FS $\tilde{A}$, i.e.,

$$\tilde{A}_e^j \equiv \left\{ \left( u_i^j, f_{x_i}\left( u_i^j \right), i = 1, \ldots, N \right) \right\}, \tag{3.6}$$

where

$$u_i^j \in \{u_{ik}, k = 1, \ldots, M_i\}. \tag{3.7}$$

Thus, we can represent $\tilde{A}$ as the union of its T2 embedded sets, i.e.,

$$\tilde{A} = \sum_{j=1}^{n} \tilde{A}_e^j \tag{3.8}$$

where

$$n \equiv \prod_{i=1}^{N} M_i \tag{3.9}$$

Figure 3.2 shows an example of vertical-slice and wavy-slice representations of the T2 FS $\tilde{A}$. For continuous universes of discourse $X$ and $U$, $\tilde{A}$ can be represented either by the union of infinite vertical-slice secondary sets (dotted line) in Fig. 3.2a, or by the union of infinite wavy-slice embedded T2 FSs (dotted line) in Fig. 3.2c. For discrete universes of discourse $X$ and $U$, consider two elements $x_1, x_2 \in X$ and $u_1, u_2 \in U$. The vertical-slice representation of $\tilde{A}$ is shown in Fig. 3.2b,

$$\tilde{A} = (0.9/0.2)/x_1 + (0.2/0.6)/x_1 + (0.5/0.8)/x_1 + (0.1/0.4)/x_2 + (0.6/0.7)/x_2. \tag{3.10}$$

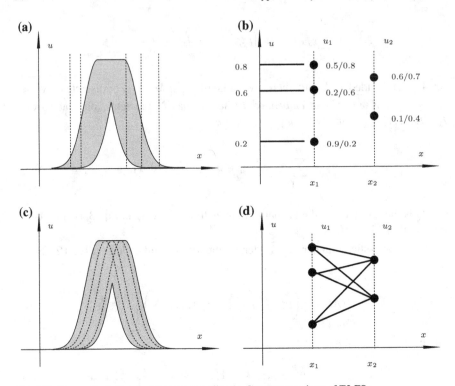

**Fig. 3.2** The vertical-slice (**a, b**) and wavy-slice (**c, d**) representations of T2 FSs

Observe that $M_1^A = 3$, $M_2^A = 2$, and $n_A = M_1^A M_2^A = 6$. Hence, there are six embedded T2 FSs, namely

$$\tilde{A}_e^1 = (0.9/0.2)/x_1 + (0.1/0.4)/x_2, \tag{3.11}$$

$$\tilde{A}_e^2 = (0.9/0.2)/x_1 + (0.6/0.7)/x_2, \tag{3.12}$$

$$\tilde{A}_e^3 = (0.2/0.6)/x_1 + (0.1/0.4)/x_2, \tag{3.13}$$

$$\tilde{A}_e^4 = (0.2/0.6)/x_1 + (0.6/0.7)/x_2, \tag{3.14}$$

$$\tilde{A}_e^5 = (0.5/0.8)/x_1 + (0.1/0.4)/x_2, \tag{3.15}$$

$$\tilde{A}_e^6 = (0.5/0.8)/x_1 + (0.6/0.7)/x_2. \tag{3.16}$$

It is easy to see that $\tilde{A} = \sum_{j=1}^6 \tilde{A}_e^j$ in the wavy-slice manner as shown in Fig. 3.2d.

The vertical-slice representation depicts the three-dimensional nature of the T2 FS without redundancy. In contrast, the wavy-slice representation contains an enormous amount of redundancy because many elements of embedded T2 FSs are duplicated. However, the concept of embedded sets bridges the T2 FSs and T1 FSs, so that the operations of T2 FSs can be viewed as the parallel operations of many embedded T1

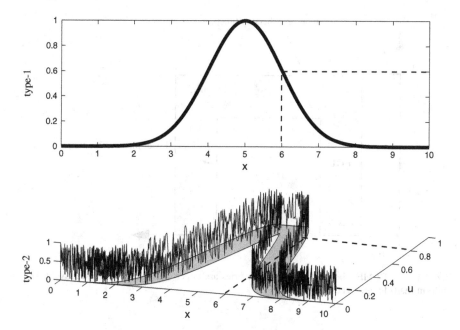

**Fig. 3.3** The T1 MF versus the T2 MF. The output of the T1 MF is a crisp number, e.g., $h_A(6) = 0.6$. The output of the T2 MF is a T1 FS, e.g., $h_{\tilde{A}}(6) = [0.2, 0.8]$, where each element is associated with a secondary grade. The FOU is a bounded uncertainty of the primary membership

FSs. Such a representation as wavy slice will be used in both theoretical derivations and practical computations.

Mathematically, the T2 MF is three-dimensional as shown in Fig. 3.1. It is the new third dimension that provides new design degrees of freedom for handling uncertainty. Two new concepts distinguish T2 MF from T1 MF: the secondary MF (the vertical slice in Fig. 3.1) and FOU (the shaded region in Fig. 3.1). The FOU is the two-dimensional domain of the T2 MF. The secondary MF sits atop of the FOU. The design of the FOU and secondary MF determines the modeling capability of the T2 MF. As shown in Fig. 3.3, the output of the T1 MF is a crisp number, e.g., $h_A(6) = 0.6$, and the output (bounded by FOU) of the T2 MF is a T1 FS, e.g., $h_{\tilde{A}}(6) = [0.2, 0.8]$, where each element is associated with a secondary grade. In mathematics, a function relates each of its inputs to exactly one output. From this view, the T2 MF is not strictly a "function" because it relates each of its inputs to a T1 FS (*one-to-many* mapping).

Through the secondary MF, we assign a precise weight to each primary membership grade in order to reflect its uncertainty. We think of these weights as the possibilities associated with embedded T1 MF at this value. The design of secondary MF usually involves prior knowledge about these possibilities. Although theoretically we can select arbitrarily any shape of secondary MF at each vertical slice of primary membership, practically at present we have difficulty in justifying

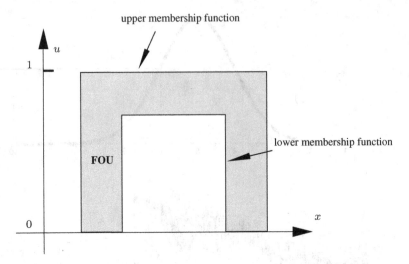

**Fig. 3.4** An IT2 MF. The two bounded boundaries are upper and lower MFs, respectively. The uniform shaded region is the FOU

the use of any kind except the uniform possibilities. The uniform possibilities over the entire FOU reduce the general T2 FS to the FS in Fig. 3.4. IT2 i.e., the secondary grades are equal to one for all primary membership grades. Therefore, we prefer to use the IT2 FS in practice by assuming that the secondary MF is the uniform MF.

Besides the secondary MF, the FOU plays an important role that retains a controlled degree of uncertainty for the primary MF. It should be pretty clear that less (more) uncertainty can be associated with a smaller (larger) FOU, but this is not very quantitative. The type-reduced set (centroid) of the IT2 FS is an interval set completely characterized by its left- and right-end points. The larger (smaller) the amount of uncertainty—as reflected by a larger (smaller) FOU—the larger (smaller) will be the centroid of the IT2 FS be. So the FOU can be measured by the centroid of the T2 FS. The design of FOU is usually concerned with the uncertain parameters of the corresponding primary MF (T1 MF).

## 3.2 Operations on Type-2 Fuzzy Sets

The T2 FS has two new operators, the *meet* (denoted by ⊓) and the *join* (denoted by ⊔), to account for the intersection and union. Consider two T2 FSs $\tilde{A}$ and $\tilde{B}$ in continuous universes of discourse $X$ and $U$. Let $h_{\tilde{A}}(x)$ and $h_{\tilde{B}}(x)$ be the membership grades (FSs in $J_x \subseteq [0, 1]$) of these two sets, represented, for each $x$, as $h_{\tilde{A}}(x) = \int_u f_x(u)/u$ and $h_{\tilde{B}}(x) = \int_w f_x(w)/w$ respectively, where $u, w \in J_x$ indicate the primary membership grades of $x$, and $f_{x(u)}, g_{x(w)} \in [0, 1]$ indicate the secondary membership grades of $x$. Using Zadeh's extension principle, the membership grades for union, intersection, and complement of T2 FSs $\tilde{A}$ and $\tilde{B}$ have been defined as follows,

$$\tilde{A} \cup \tilde{B} \Leftrightarrow h_{\tilde{A} \cup \tilde{B}}(x) = h_{\tilde{A}}(x) \sqcup h_{\tilde{B}}(x) = \int_u \int_w (f_x(u) \star g_x(w))/(u \vee w), \ x \in X, \tag{3.17}$$

$$\tilde{A} \cap \tilde{B} \Leftrightarrow h_{\tilde{A} \cap \tilde{B}}(x) = h_{\tilde{A}}(x) \sqcap h_{\tilde{B}}(x) = \int_u \int_w (f_x(u) \star g_x(w))/(u \star w), \ x \in X, \tag{3.18}$$

$$\bar{\tilde{A}} \Leftrightarrow h_{\bar{\tilde{A}}} = \neg h_{\tilde{A}}(x) = \int_u f_x(u)/(1 - u), \ x \in X. \tag{3.19}$$

For discrete universes of discourse $X$ and $U$, the wavy-slice representation of the union is

$$\tilde{A} \cup \tilde{B} = \sum_{j=1}^{n_A} \sum_{i=1}^{n_B} \tilde{A}_e^j \cap \tilde{B}_e^i$$

$$= \sum_{j=1}^{n_A} \sum_{i=1}^{n_B} \left\{ \left[ f_{x_1}(u_1^j) \star g_{x_1}(w_1^i) \Big/ u_1^j \vee w_1^i \right] \Big/ x_1 + \ldots \right.$$

$$\left. + \left[ f_{x_N}(u_N^j) \star g_{x_N}(w_I^i) \Big/ u_N^j \vee w_N^i \right] \Big/ x_N \right\}. \tag{3.20}$$

where $\star$ is a $t$-norm (e.g., minimum, product, etc.), and $\vee$ is the *maximum* $t$-conorm. Equation (3.20) can also be expressed as vertical-slice representation,

$$\tilde{A} \cup \tilde{B} = \sum_{j=1}^{M_1(u_1)} \sum_{i=1}^{M_1(w_1)} \left[ f_{x_1}(u_1^j) \star g_{x_1}(w_1^i) \Big/ u_1^j \vee w_1^i \right] \Big/ x_1 + \ldots$$

$$+ \sum_{j=1}^{M_N(u_N)} \sum_{i=1}^{M_N(w_N)} \left[ f_{x_N}(u_N^j) \star g_{x_N}(w_I^i) \Big/ u_N^j \vee w_N^i \right] \Big/ x_N. \tag{3.21}$$

The wavy-slice representation of the intersection is

$$\tilde{A} \cup \tilde{B} = \sum_{j=1}^{n_A} \sum_{i=1}^{n_B} \left\{ \left[ f_{x_1}(u_1^j) \star g_{x_1}(w_1^i) \Big/ u_1^j \wedge w_1^i \right] \Big/ x_1 + \ldots \right.$$

$$\left. + \left[ f_{x_N}(u_N^j) \star g_{x_N}(w_I^i) \Big/ u_N^j \wedge w_N^i \right] \Big/ x_N \right\}. \tag{3.22}$$

Equation (3.22) can also be expressed as vertical-slice representation,

$$
\tilde{A} \cup \tilde{B} = \sum_{j=1}^{M_1(u_1)} \sum_{i=1}^{M_1(w_1)} \left[ f_{x_1}(u_1^j) \star g_{x_1}(w_1^i) \Big/ u_1^j \wedge w_1^i \right] \Big/ x_1 + \dots
$$

$$
+ \sum_{j=1}^{M_N(u_N)} \sum_{i=1}^{M_N(w_N)} \left[ f_{x_N}(u_N^j) \star g_{x_N}(w_I^i) \Big/ u_N^j \wedge w_N^i \right] \Big/ x_N.
$$

$$(3.23)$$

The wavy-slice representation of the complement is

$$
\bar{\tilde{A}} = \sum_{j=1}^{n_A} \left( \sum_{i=1}^{N} \left[ f_{x_i}(u_i^j) \Big/ (1 - u_i^j) \right] \Big/ x_i \right),
$$

$$(3.24)$$

where $n_A$ is given by Eq. (3.9). Equation (3.24) can also be expressed as vertical-slice representation,

$$
\bar{\tilde{A}} = \sum_{j=1}^{N} \left( \sum_{j=1}^{M_i} \left[ f_{x_i}(u_i^j) \Big/ (1 - u_i^j) \right] \Big/ x_i \right).
$$

$$(3.25)$$

The IT2 FSs are the most widely used T2 FSs because they are simple to use and because, at present, it is very difficult to justify the use of any other kind. The IT2 FSs have all secondary grades equal to one as shown in Fig. 7.18. In this case, we treat embedded T2 FSs as embedded T1 FSs so that no new concepts are needed to derive the union, intersection, and complement of such sets. After each derivation, we merely append interval secondary grades to all the results in order to obtain the final formulas for the union, intersection, and complement of IT2 FSs, e.g., in Eqs. (3.17)–(3.24), $\forall f_x(u) = 1$, $\forall g_x(w) = 1$, and $\forall f_x(u) \star g_x(w) = 1$.

The operations on the IT2 FS,

$$
h_{\tilde{A}(x)} = \int_{u \in [\underline{h}_{\tilde{A}}(x), \overline{h}_{\tilde{A}}(x)]} 1/u,
$$

$$(3.26)$$

can be also represented by its lower and upper MFs $\underline{h}_{\tilde{A}}(x)$ and $\overline{h}_{\tilde{A}}(x)$. The *meet, join* and *negation* operations of interval sets are derived from Eqs. (3.17)–(3.19) simply by replacing $f_x(u)$ and $g_x(u)$ to 1,

$$
h_{\tilde{A}}(x) \sqcup h_{\tilde{B}}(x) = \int_{u \vee w \in [\underline{h}_{\tilde{A}}(x) \vee \underline{h}_{\tilde{B}}(x), \overline{h}_{\tilde{A}}(x) \vee \overline{h}_{\tilde{A}}(x)]} 1 \Big/ u \vee w,
$$

$$(3.27)$$

$$h_{\tilde{A}}(x) \sqcap h_{\tilde{B}}(x) = \int_{u \star w \in [\underline{h}_{\tilde{A}}(x) \star \underline{h}_{\tilde{B}}(x), \overline{h}_{\tilde{A}}(x) \star \overline{h}_{\tilde{A}}(x)]} 1 \Big/ u \star w, \qquad (3.28)$$

$$\neg h_{\tilde{A}}(x) = \int_{u \in [\underline{h}_{\tilde{A}}(x), \overline{h}_{\tilde{A}}(x)]} 1 \Big/ (1 - u). \qquad (3.29)$$

## 3.3 Type-2 Fuzzy Logic Systems

Figure 3.5 shows the structure of a T2 fuzzy logic system (FLS). The general T2 FLS includes five basic parts: fuzzifier, rule base, fuzzy inference engine, type reducer, and defuzzifier. The fuzzifier maps a crisp input into a FS, which can be T1 or T2 FSs. In the "IF-THEN" rule base, as long as one antecedent or the consequent set is T2 FS, we will have a T2 FLS. The inference engine combines rules and gives a mapping from input T2 FSs to output T2 FSs, in which unions and intersections of T2 FSs are used, as well as composition of T2 relations. The type reducer produces a T1 FS from the output T2 FS of the inference engine. To further obtain a crisp output, we defuzzify the type-reduced set. In contrast to the T1 FLS, besides we replace all T1 FSs as well as operations by T2 FSs, we have an additional type reducer mapping the output T2 FS into T1 FS. From the embedded T1 FSs' perspective, the T2 FLS is a collection of all the embedded T1 FLSs.

### 3.3.1 Fuzzifier

The fuzzifier maps a crisp point $x \in U$ into a fuzzy set $X \in U$. In the case of a singleton fuzzifier, the crisp point $X \in U$ is mapped into a fuzzy set $X$ with support $x_i$, where $h_X(x_i) = 1$ for $x_i = x$ and $h_X(x_i) = 0$ for $x_i \neq x$, i.e., the single point

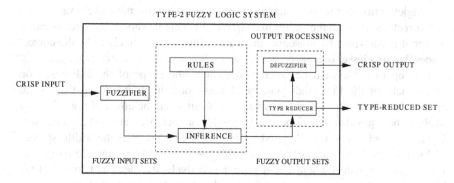

**Fig. 3.5** The structure of a T2 FLS

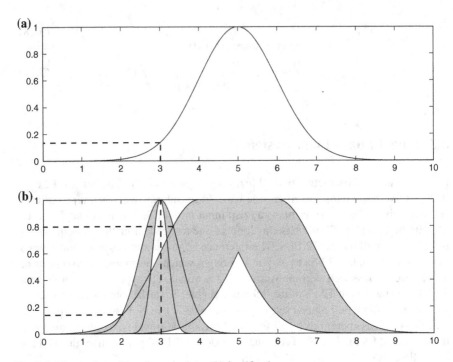

**Fig. 3.6** The singleton (a) and nonsingleton (b) fuzzification

in the support of with nonzero MF value is $x_i = x$. In the case of a nonsingleton fuzzifier, the point $X \in U$ is mapped into a FS $X$ with support $x_i$, where $h_X$ achieves maximum value at $x_i = x$ and decreases while moving away from $x_i = x$. We assume that FS $X$ is normalized so that $h_X(x) = 1$. Figure 3.6 shows the singleton fuzzification and T2 fuzzy nonsingleton fuzzification.

The nonsingleton fuzzification is especially useful in cases where the available training data, or the input data to the FLS, are corrupted by noise. Conceptually, the nonsingleton fuzzifier implies that the given input value is the most likely value to be the correct one from all the values in its immediate neighborhood; however, because the input is corrupted by noise, neighboring points are also likely to be the correct values, but to a lesser degree.

It is up to the system designer to determine the shape of the MF based on an estimate of the kind and quantity of noise present. It would be the logical choice, though, for the MF to be symmetric about since the effect of noise is most likely to be equivalent on all points. Examples of such MFs are: (1) the Gaussian $h_X(x_i) = exp[-(x - x_i)^2/2\sigma^2]$ where the variance $\sigma^2$ reflects the width (spread) of $h_X(x_i)$, (2) triangular $h_X(x_i) = \max(0, 1 - |(x - x_i)/c|)$, and (3) $h_X(x_i) = 1/(1 + |(x - x_i)/c|^p)$ where $x$ and $c$ are, respectively, the mean and spread of the FSs. Note that larger values of the spread of the above MFs imply that more noise is anticipated to exist in the given data. Greater uncertainty in the input not only "fires" rules at a higher level than a singleton system, but it also usually fires more rules than a singleton system would. This is due to the fact that a nonsingleton input may have membership in more antecedent fuzzy regions than a singleton input.

If singleton fuzzifier is used, the corresponding T2 FLS is called singleton T2 FLS (ST2 FLS). If inputs are modeled as T1 or T2 fuzzy numbers, it is called T1 or T2 nonsingleton fuzzification, and the corresponding T2 FLS is called T1 or T2 nonsingleton T2 FLS (T1NT2 or T2NT2 FLS).

### 3.3.2 Rule Base and Inference

Consider a T2 FLS having $p$ inputs, $x_1 \in X_1, x_2 \in X_2, \ldots, x_p \in X_p$, and one output $y \in Y$. Let us suppose that it has $M$ rules where the $l$th rule has the form,

$$R^l : \text{IF } x_1 \text{ is } \tilde{F}_1^l \text{ and } x_2 \text{ is } \tilde{F}_2^l \text{ and } \ldots \text{and } x_p \text{ is } \tilde{F}_p^l, \text{ THEN } y \text{ is } \tilde{G}^l. \quad (3.30)$$

This rule represents a T2 fuzzy relation between the input space, $X_1 \times X_2 \times \cdots \times X_p$, and the output space $Y$ of the FLS.

Based on rule base Eq.(3.30), we denote the MF of this T2 relation as $h_{\tilde{F}_1^l \times \cdots \times \tilde{F}_p^l \to \tilde{G}^l}$

$(\mathbf{x}, y)$, where $\tilde{F}_1^l \times \cdots \times \tilde{F}_p^l$ denotes the Cartesian product of $\tilde{F}_1^l, \tilde{F}_2^l, \ldots, \tilde{F}_p^l$, and $\mathbf{x} = \{x_1, x_2, \ldots, x_p\}$. The composition of the FS $\tilde{X}$ and the rule $R^l$ is found by using the extended sup-star composition,

$$h_{\tilde{X} \circ \tilde{F}_1^l \times \cdots \times \tilde{F}_p^l \to \tilde{G}^l}(y) = \sqcup_{x \in \tilde{X}} [h_{\tilde{X}}(x) \sqcap h_{\tilde{F}_1^l \times \cdots \times \tilde{F}_p^l \to \tilde{G}^l}(x, y)]. \quad (3.31)$$

If we denote $\tilde{X} \circ \tilde{F}_1^l \times \cdots \times \tilde{F}_p^l \to \tilde{G}^l(y)$ by $\tilde{B}^l$, in the ST2 FLS, we can rewrite Eq.(3.31) as follows,

$$h_{\tilde{B}^l}(y) = h_{\tilde{G}^l}(y) \sqcap \left[ \sqcap_{i=1}^p h_{\tilde{F}_i^l}(x_i) \right]. \quad (3.32)$$

In the T1NT2 FLS, we can rewrite Eq.(3.31) as follows,

$$h_{\tilde{B}^l}(y) = h_{\tilde{G}^l}(y) \sqcap \left\{ \sqcap_{i=1}^p \left[ \sqcup_{x_i \in X_i} \left( h_{X_i}(x_i) \sqcap h_{\tilde{F}_i^l}(x_i) \right) \right] \right\}. \quad (3.33)$$

In the T2NT2 FLS, we can rewrite Eq.(3.31) as follows,

$$h_{\tilde{B}^l}(y) = h_{\tilde{G}^l}(y) \sqcap \left\{ \sqcap_{i=1}^p \left[ \sqcup_{x_i \in X_i} \left( h_{\tilde{X}_i}(x_i) \sqcap h_{\tilde{F}_i^l}(x_i) \right) \right] \right\}. \quad (3.34)$$

A general T2 FLS is complicated because the computation is prohibitive in the set operation. Moreover, we have no rational basis for choosing secondary MFs. The IT2 FS is simple to use because the above operations are concerned only with the upper and lower MFs. Thus in an IT2 FLS, we denote $t$-norm by $\mathcal{T}$ and $h_{\tilde{B}^l}(y) = [\underline{h}_{\tilde{B}^l}(y), \overline{h}_{\tilde{B}^l}(y)]$. In the SIT2 FLS, from Eq. (3.32) we have

$$\underline{h}_{\tilde{B}^l}(y) = \underline{h}_{\tilde{G}^l}(y) \star \mathcal{T}_{i=1}^{p} \underline{h}_{\tilde{F}_i^l}(x_i), \tag{3.35}$$

$$\overline{h}_{\tilde{B}^l}(y) = \overline{h}_{\tilde{G}^l}(y) \star \mathcal{T}_{i=1}^{p} \overline{h}_{\tilde{F}_i^l}(x_i). \tag{3.36}$$

In the T1NIT2 FLS, from Eq. (3.33) we have

$$\underline{h}_{\tilde{B}^l}(y) = \underline{h}_{\tilde{G}^l}(y) \star \left\{ \mathcal{T}_{i=1}^{p} \left[ \sup_{x_i \in X_i} \left( h_{X_i}(x_i) \star \underline{h}_{\tilde{F}_i^l}(x_i) \right) \right] \right\}, \tag{3.37}$$

$$\overline{h}_{\tilde{B}^l}(y) = \overline{h}_{\tilde{G}^l}(y) \star \left\{ \mathcal{T}_{i=1}^{p} \left[ \sup_{x_i \in X_i} \left( h_{X_i}(x_i) \star \overline{h}_{\tilde{F}_i^l}(x_i) \right) \right] \right\}. \tag{3.38}$$

In the IT2NIT2 FLS, from Eq. (3.34) we have

$$\underline{h}_{\tilde{B}^l}(y) = \underline{h}_{\tilde{G}^l}(y) \star \left\{ \mathcal{T}_{i=1}^{p} \left[ \sup_{x_i \in X_i} \left( \underline{h}_{\tilde{X}_i}(x_i) \star \underline{h}_{\tilde{F}_i^l}(x_i) \right) \right] \right\}, \tag{3.39}$$

$$\overline{h}_{\tilde{B}^l}(y) = \overline{h}_{\tilde{G}^l}(y) \star \left\{ \mathcal{T}_{i=1}^{p} \left[ \sup_{x_i \in X_i} \left( \overline{h}_{\tilde{X}_i}(x_i) \star \overline{h}_{\tilde{F}_i^l}(x_i) \right) \right] \right\}. \tag{3.40}$$

### 3.3.3 Type Reducer and Defuzzifier

The output set corresponding to each rule in Eq. (3.30) of the T2 FLS is a T2 FS Eq. (3.31). The type reducer combines all these output sets in some way and then performs a centroid calculation on this T2 FS, which leads to a T1 set that we call the type-reduced set. Hence the type reducer is a mapping from the T2 FS to the T1 FS representing T2 FS's centroid. According to different combination methods of output T2 FSs, we mainly have three type-reduction methods: centroid, height, and center-of-sets type reducers.

Similar to the centroid of the T1 FS, the centroid of a T2 set $\tilde{A}$, whose domain is discretized into $N$ points, can be defined using the extension principle as follows. Let $J_{x_i} = h_{\tilde{A}}(x_i)$, so that

$$C_{\tilde{A}} = \int_{u_1} \cdots \int_{u_N} [f_{J_{x_1}}(u_1) \star \cdots \star f_{J_{x_N}}(u_N)] \bigg/ \frac{\sum_{i=1}^{N} x_i u_i}{\sum_{i=1}^{N} u_i}, \tag{3.41}$$

where $u_i \in J_{x_i}$. Equation (3.41) can be described in terms of embedded T1 sets Eq. (3.5). The centroid $C_{\tilde{A}}$ is a T1 FS whose elements are the centroids of all the embedded T1 sets in $\tilde{A}$. The membership grade of the element of $C_{\tilde{A}}$ is calculated as the $t$-norm of the secondary grades $\{f_{J_{x_1}}(u_1), \ldots, f_{J_{x_N}}(u_N)\}$ that make up that embedded set.

Observe that the number of embedded T1 sets is uncountable in the continuous discourses of universe $X$ and $U$. Therefore, the domain of $\tilde{A}$ and $J_{x_i}$ have to be discretized for the calculation $C_{\tilde{A}}$. Observe from Eq. (3.41) that if the domain of each $J_{x_i}$ is discretized into $M$ points, the number of possible $\{f_{J_{x_1}}(u_1), \ldots, f_{J_{x_N}}(u_N)\}$ combinations is $M^N$, which can be very large even for small $M$ and $N$. In the following subsections, we present details for centroid, hight, and center-of-sets type-reduction. Computational complexity of different type reducers are discussed. In particular, we give the exact centroid of an IT2 FS by the Karnik-Mendel (KM) algorithm.

### 3.3.3.1 Centroid Type Reducer

The centroid type reducer combines all the output T2 sets by finding their union. The membership grade of $y \in Y$ is given as

$$h_{\tilde{B}}(y) = \sqcup_{l=1}^{M} h_{\tilde{B}^l}(y), \tag{3.42}$$

where $h_{\tilde{B}}(y)$ is as defined in Eq. (3.31). The centroid type reducer then calculates the centroid of $\tilde{B}$,

$$Y_c(x) = \int_{u_1} \cdots \int_{u_N} [h_{J_1}(u_1) \star \cdots \star h_{J_N}(u_N)] \bigg/ \frac{\sum_{i=1}^{N} y_i u_i}{\sum_{i=1}^{N} u_i}, \tag{3.43}$$

where $u_i \in J_i = h_{\tilde{B}}(y_i)$, $1 \le i \le N$.

The sequence of computations needed to obtain $Y_c(x)$ is as follows.

- Compute the combined output set using Eq. (3.42).
- Discretize the output space $Y$ into $N$ points, $y_1, \ldots, y_N$.
- Discretize the domain of each $h_{\tilde{B}}(y_i)$ into $M_i$ points.
- Enumerate all the $\prod_{j=1}^{N} M_j$ embedded sets.
- Compute the centroid type-reduced set using Eq. (3.43). We must use the minimum $t$-norm here.

### 3.3.3.2 Height Type Reducer

For $l$th rule in Eqs. (3.32)–(3.34), the $l$th output T2 FSs may have a vertical slice at point $\bar{y}^l$, where $\bar{y}^l$ can be chosen to be the point having the highest membership in the principal MF of output set $\tilde{B}^l$. If $\tilde{B}^l$ is such that a principal MF cannot be defined, one may choose $\bar{y}^l$ as the point having the highest primary membership with a secondary membership equal to one or as a point satisfying some similar criterion. All vertical slices from the total number of $M$ output T2 FSs form a new combined T2 FS,

$$h_{\tilde{B}}(\bar{y}) = \sum_{l=1}^{M} h_{\tilde{B}^l}(\bar{y}^l), \qquad (3.44)$$

where for example, in ST2 FLS,

$$h_{\tilde{B}^l}(\bar{y}^l) = h_{\tilde{G}^l}(\bar{y}^l) \sqcap [\sqcap_{i=1}^{p} h_{\tilde{F}_i^l}(x^i)]. \qquad (3.45)$$

If we let $J^l = h_{\tilde{B}^l}(\bar{y}^l)$, the expression for the type-reduced set is obtained as

$$Y_h(x) = \int_{u_1} \cdots \int_{u_M} [h_{J^1}(u_1) \star \cdots \star h_{J^M}(u_M)] \bigg/ \frac{\sum_{l=1}^{M} \bar{y}_i u_i}{\sum_{l=1}^{M} u_i}, \qquad (3.46)$$

where $u_l \in J^l, 1 \le l \le M$.

The sequence of computations needed to obtain $Y_h(x)$ is as follows.

- Choose $\bar{y}^l$ for each output set, $l = 1, 2, \ldots, M$, and compute $h_{\tilde{B}^l}(\bar{y}^l)$ such as in Eq. (3.45).
- Discretize the domain of each $h_{\tilde{B}^l}(\bar{y}^l)$ into $N_l$ points.
- Enumerate all the possible combinations $\{u_1, \ldots, u_M\}$ such that $u_l \in h_{\tilde{B}^l}(\bar{y}^l)$. There will be $\prod_{l=1}^{M} N_l$ combinations.
- Compute the height type-reduced set using Eq. (3.46).

### 3.3.3.3 Center-of-Sets Type Reducer

The center-of-sets type reducer replaces each consequent set by its centroid and finds a weighted average of these centroids, where the weight associated with the $l$th centroid is the degree of firing corresponding to the $l$th rule, namely $\sqcap_{i=1}^{p} h_{\tilde{F}_i^l}(x_i)$ as in the ST2 FLS Eq. (3.32). The center-of-sets type reducer combines the output T2 FSs from all rules into a new T2 FS,

$$h_{\tilde{B}}(d_l) = \sum_{l=1}^{M} [h_{C_l}(d_l) \star h_{E_l}(e_l)/e_l]/d_l, \qquad (3.47)$$

where $\star$ indicates the chosen $t$-norm; $d_l \in C_l = C_{\tilde{G}^l}$ the centroid of the $l$th consequent set and $e_l \in E_l = \sqcap_{i=1}^{p} \mu_{\tilde{F}_i^l}(x_i)$, the degree of firing associated with the $l$th consequent set for $1 \le l \le M$. The expression for the type-reduced set is

$$Y_{cos}(x) = \int_{d_1} \cdots \int_{d_M} \int_{e_1} \cdots \int_{e_M} \mathcal{T}_{l=1}^{M} \mu_{C_l}(d_l) \star \mathcal{T}_{l=1}^{M} \mu_{E_l}(e_l) \bigg/ \frac{\sum_{l=1}^{M} d_l e_l}{\sum_{l=1}^{M} e_l}, \qquad (3.48)$$

where $\mathcal{T}$ indicates the chosen $t$-norm.

The sequence of computations needed to obtain $Y_{cos}(x)$ is as follows.

- Discretize the output space $Y$ into a number of points and compute the centroid $C_{\tilde{G}^l}$ of each consequent set on the discretized output space using Eq. (3.41).
- Compute the degree of firing $E_l = \sqcap_{i=1}^{p} \mu_{\tilde{F}_i^l}(x_i)$ associated with the $l$th consequent set, where $1 \leq l \leq M$.
- Discretize the domain of each $C_{\tilde{G}^l}$ into $M_l$ points, where $1 \leq l \leq M$.
- Discretize the domain of each $E_l$ into $N_l$ points, where $1 \leq l \leq M$.
- Enumerate all the possible combinations, $\{d_1, \ldots, d_M, e_1, \ldots, e_M\}$, such that $d_l \in C_{\tilde{G}^l}$ and $e_l \in E_l$. The total number of combinations will be $\prod_{j=1}^{M} M_j N_j$.
- Compute the type-reduced set using Eq. (3.48).

### 3.3.3.4 The Karnik-Mendel Algorithm

Computing the centroid of a general T2 FS involves an enormous amount of computation; however, if the MFs of the primary membership have a regular structure, we can obtain the exact or approximate centroid without having to do all the calculations. Especially, computing the centroid of an IT2 FS only involves two independent iterative procedures that can be performed in parallel, namely the Karnik-Mendel (KM) algorithm. Because the centroid of each of the embedded T1 FSs is a bounded number, the centroid of the IT2 FS is an interval set,

$$C_{\tilde{A}} = 1/[c_l, c_r], \tag{3.49}$$

$$c_l = \min_{\forall u_i \in [\underline{h}_{\tilde{A}}(x_i), \overline{h}_{\tilde{A}}(x_i)]} \sum_{i=1}^{N} x_i u_i \bigg/ \sum_{i=1}^{N} u_i, \tag{3.50}$$

$$c_r = \max_{\forall u_i \in [\underline{h}_{\tilde{A}}(x_i), \overline{h}_{\tilde{A}}(x_i)]} \sum_{i=1}^{N} x_i u_i \bigg/ \sum_{i=1}^{N} u_i. \tag{3.51}$$

The discrete version of the KM algorithm computes $c_l$ and $c_r$ by at most $N$ iterations as shown in Fig. 3.7. Figure 3.8 shows the continuous version of the KM algorithm that converges monotonically. By the KM algorithm, we can easily obtain the centroids of the output IT2 FSs in Eqs. (3.42), (3.45), and (3.47) in order to realize the centroid, height, and center-of-set type-reductions.

### 3.3.3.5 Defuzzifier

We defuzzify the type-reduced set to get a crisp output from the T2 FLS. The most natural way of doing this seems to be by finding the centroid of the type-reduced set. Finding the centroid is equivalent to finding a weighted average of the outputs of all the T1 FLS's embedded in the T2 FLS, where the weights correspond to the memberships in the type-reduced set. If the type-reduced set $Y$ for an input is discrete

$$
\begin{array}{ll}
\textbf{input} & : \tilde{A} = [\,\underline{h}(x_i),\,\overline{h}(x_i)\,],\ 1 \le i \le N. \\
\textbf{output} & : C_{\tilde{A}} = [\,c_l, c_r\,]. \\
\textbf{initialize:} & u_i \leftarrow \frac{1}{2}[\underline{h}(x_i) + \overline{h}(x_i)],\ 1 \le i \le N.
\end{array}
$$

1  **begin**
2　　$c_l, c_r \leftarrow \sum_{i=1}^{N} x_i u_i / \sum_{i=1}^{N} u_i;$
3　　**repeat**
4　　　　$c_l' \leftarrow c_l, c_r' \leftarrow c_r;$
5　　　　$k_l \leftarrow \arg_k x_k \le c_l \le x_{k+1}, k\ \ r \leftarrow \arg_k x_k \le c_r \le x_{k+1};$
6　　　　$c_l \leftarrow$
　　　　$[\sum_{i=1}^{k_l} x_i \overline{h}(x_i) + \sum_{i=k_l+1}^{N} x_i \underline{h}(x_i)] / [\sum_{i=1}^{k_l} \overline{h}(x_i) + \sum_{i=k_l+1}^{N} \underline{h}(x_i)];$
7　　　　$c_r \leftarrow$
　　　　$[\sum_{i=1}^{k_r} x_i h(x_i) + \sum_{i=k_r+1}^{N} x_i \overline{h}(x_i)] / [\sum_{i=1}^{k_r} \underline{h}(x_i) + \sum_{i=k_r+1}^{N} \overline{h}(x_i)];$
8　　**until** $c_l = c_l', c_r = c_r'$ ;
9  **end**

**Fig. 3.7** The Karnik-Mendel algorithm (discrete version)

$$
\begin{array}{ll}
\textbf{input} & : \tilde{A} = [\,\underline{h}(x),\,\overline{h}(x)\,]. \\
\textbf{output} & : C_{\tilde{A}} = [\,c_l, c_r\,]. \\
\textbf{initialize:} & \alpha_l, \alpha_r \leftarrow \int_{-\infty}^{\infty} x[\overline{h}(x) + \underline{h}(x)]\, dx / \int_{-\infty}^{\infty} [\overline{h}(x) + \underline{h}(x)]\, dx.
\end{array}
$$

1  **begin**
2　　**repeat**
3　　　　$\alpha_l \leftarrow c_l, \alpha_r \leftarrow c_r;$
4　　　　$c_l \leftarrow [\int_{-\infty}^{\alpha_l} x\overline{h}(x)\, dx + \int_{\alpha_l}^{\infty} x\underline{h}(x)\, dx] / [\int_{-\infty}^{\alpha_l} \overline{h}(x)\, dx + \int_{\alpha_l}^{\infty} \underline{h}(x)\, dx];$
5　　　　$c_r \leftarrow [\int_{-\infty}^{\alpha_r} x\underline{h}(x)\, dx + \int_{\alpha_r}^{\infty} x\overline{h}(x)\, dx] / [\int_{-\infty}^{\alpha_r} \underline{h}(x)\, dx + \int_{\alpha_r}^{\infty} \overline{h}(x)\, dx];$
6　　**until** $\alpha_l = c_l,\ \alpha_r = c_r$ ;
7  **end**

**Fig. 3.8** The Karnik-Mendel algorithm (continuous version)

and consists of $N$ points, the expression for its centroid is

$$
C_Y(x) = \frac{\sum_{k=1}^{N} y_k h_Y(y_k)}{\sum_{k=1}^{N} h_Y(y_k)}. \tag{3.52}
$$

We use $y = \mathfrak{D}(\tilde{A})$ to denote both type reducer and defuzzifier of the T2 FS $\tilde{A}$, which is a mapping from the T2 FS to a crisp number.

## 3.4 Pattern Recognition Using Type-2 Fuzzy Sets

Pattern recognition typically involves the classification of unknown observation data **x** (pattern) to the given class model $\lambda_\omega$, $1 \le \omega \le C$. Let $\kappa(\mathbf{x}, \lambda_\omega)$ be a similarity measure that measures the degree of similarity, or compatibility, between the unknown pattern **x** and the $\omega$th class model $\lambda_\omega$. Thus, we may formally write the process of classifying the unknown pattern **x** to the class $\omega^*$, where

$$\omega^* = \arg \max_{\omega=1}^{C} \kappa(\mathbf{x}, \lambda_\omega). \tag{3.53}$$

In Sect. 2.3, the similarity measure is the posterior probability Eq. (2.4) according to the Bayesian decision rule. However, uncertainties exist in both *data* and class *model*. Data uncertainty is uncertainty in the input variables, such as noise or incomplete information. For example, speech data have the following uncertainties: (1) The same phoneme has different values in different contexts; (2) the same phoneme has different lengths or different frames; (3) the beginning and the end of a phoneme are uncertain. A model may be interpreted as a set of elements and rules that map input variables onto output variables. Model uncertainty is uncertainty in the mapping induced by uncertain parameters in the model. The reason may be insufficient training data or incomplete information. Therefore, the probability measure in Eq. (2.4) is not suitable for the similarity measure between the unknown pattern and class model, because it is insufficient to describe both data and model uncertainty.

The T2 MF has a three-dimensional structure in which the secondary MF and FOU describe the uncertainty of the primary MF. We may use one dimension, the primary MF, as the class model $\lambda_\omega$, and use another dimension, the secondary MF, to evaluate the uncertainty of the primary MF. So the model uncertainty is described by the T2 FS $\tilde{\lambda}_\omega$. The fuzzification describes the uncertainty of input variables as mentioned in Sect. 3.3.1. The singleton fuzzification assumes no uncertainty in the input data space. The T2 (T1) nonsingleton fuzzification assumes the input variable as a T2 (T1) FS. Thus, the data uncertainty can be modeled by the T2 nonsingleton fuzzification $\tilde{\mathbf{x}}$. Given $\tilde{\lambda}_\omega$ and $\tilde{\mathbf{x}}$, we may rewrite Eq. (3.53) by the similarity measure between these two T2 FSs,

$$\omega^* = \arg \max_{\omega=1}^{C} \tilde{\kappa}(\tilde{\mathbf{x}}, \tilde{\lambda}_\omega). \tag{3.54}$$

Here we use the primary MF to characterize the randomness of data, and use the secondary MF to evaluate the fuzziness of the primary MF. Hence both randomness and fuzziness can be accounted for within the T2 FS framework. The main advantage of using T2 similarity measure is that T2 FSs incorporate both data and model uncertainties by the proper secondary MF and FOU. The main disadvantage is the intractable computational complexity induced by set operations.

Equation (3.54) shows how to use T2 FSs for handling uncertainty in problems of pattern recognition. We can compute the T2 similarity measure $\tilde{\kappa}$ in many ways. One is based on the wavy-slice representation of T2 FS. We sum the similarity measure of all embedded T1 FSs with different weights, i.e.,

$$\tilde{\kappa}(\tilde{A}, \tilde{B}) = \sum_{m=1}^{M} \sum_{n=1}^{N} \kappa(A_e^m, B_e^n) w_{mn}, \tag{3.55}$$

where $\kappa$ is the T1 similarity measure, $w_{mn}$ is the weight based on the secondary with the grade $m$th and $n$th embedded T1 set, and there are totally $M$, $N$ embedded T1 sets in $\tilde{A}$, $\tilde{B}$ respectively.

Also we can design classifiers based on the FLS. Consider unknown pattern $\mathbf{x} = (x_1, \ldots, x_d)'$, and two class models $\lambda_1$ and $\lambda_2$. For T1 fuzzy classifiers with a rule base of $M$ rules, each having $d$ antecedents, the $l$th rule, $R^l$, $1 \le l \le M$, is

$$R^l : \text{IF } x_1 \text{ is } F_1^l \text{ and } x_2 \text{ is } F_2^l \text{ and } \ldots \text{and } x_d \text{ is } F_d^l,$$
$$\text{THEN } \mathbf{x} \text{ is classified to } \lambda_1 \ (+1) \text{ [or is classified to } \lambda_2 \ (-1)]. \tag{3.56}$$

Suppose that the antecedents $F_i^l$, $1 \le i \le I$ are described by a T1 Gaussian MF,

$$h_{F_i^l}(x_i) = \exp\left[-\frac{1}{2}\left(\frac{x_i - \mu_i}{\sigma_i}\right)^2\right]. \tag{3.57}$$

We use the unnormalized output for the T1 FLS (the firing strength of each rule is denoted by $f^l$), namely

$$y = \sum_{l=1}^{M}(f_{\lambda_1}^l - f_{\lambda_2}^l), \tag{3.58}$$

because we make a decision based on the sign of the output ($y > 0$, $\mathbf{x} \to \lambda_1$), and normalization operation will not change the sign. For T2 fuzzy classifiers with a rule base of $M$ rules, the $l$th rule, $R^l$, $1 \le l \le M$, is

$$R^l : \text{IF } \tilde{x}_1 \text{ is } \tilde{F}_1^l \text{ and } \tilde{x}_2 \text{ is } \tilde{F}_2^l \text{ and } \ldots \text{and } \tilde{x}_d \text{ is } \tilde{F}_d^l,$$
$$\text{THEN } \tilde{\mathbf{x}} \text{ is classified to } \tilde{\lambda}_1 \ (+1) \text{ [or is classified to } \tilde{\lambda}_2 \ (-1)]. \tag{3.59}$$

Suppose that the antecedents $\tilde{F}_i^l$, $1 \le i \le I$ are described by a T2 Gaussian primary MF with uncertain mean or standard deviation. Similarly, we rewrite Eq. (3.58) for the output of the T2 FLS,

$$\tilde{y} = \sqcup_{l=1}^{M}(\tilde{f}_{\lambda_1}^l - \tilde{f}_{\lambda_2}^l). \tag{3.60}$$

The Bayesian decision theory provides the optimal solution to the general decision-making problem. If equal prior class probability is assumed, the Bayesian classifiers are

$$p(\mathbf{x}|\lambda_1) = \sum_{l=1}^{m} p(\mathbf{x}|\lambda_1^l), \tag{3.61}$$

$$p(\mathbf{x}|\lambda_2) = \sum_{l=1}^{n} p(\mathbf{x}|\lambda_2^l), \tag{3.62}$$

where we assume the number of prototypes of class $\lambda_1$ and $\lambda_2$ is $m$ and $n$, respectively. If we describe the conditional probability of each prototype by the Gaussian distribution,

$$p(\mathbf{x}|\lambda) = \frac{1}{\sqrt{(2\pi)^d |\mathbf{\Sigma}|}} e^{-\frac{1}{2}(\mathbf{x}-\boldsymbol{\mu})'\mathbf{\Sigma}^{-1}(\mathbf{x}-\boldsymbol{\mu})}, \tag{3.63}$$

where the corresponding mean vector, $\boldsymbol{\mu} = (\mu_1, \mu_2, \ldots, \mu_d)'$, and covariance matrix, $\mathbf{\Sigma} = diag(\sigma_1^2, \sigma_2^2, \ldots, \sigma_d^2)$. Based on the Bayesian decision theory, we obtain the decision rule,

$$\text{IF } p(\mathbf{x}|\lambda_1) - p(\mathbf{x}|\lambda_2) > 0, \text{ THEN } \mathbf{x} \text{ is classified to } \lambda_1, \tag{3.64}$$

$$\text{IF } p(\mathbf{x}|\lambda_1) - p(\mathbf{x}|\lambda_2) < 0, \text{ THEN } \mathbf{x} \text{ is classified to } \lambda_2. \tag{3.65}$$

Usually, we have the relationship between the number of rules in T1 fuzzy classifiers and the number of prototypes of class $\lambda_1$ and $\lambda_2$ in Bayesian classifiers,

$$M = m + n. \tag{3.66}$$

Observing Eqs. (3.58), (3.64), and (3.65), we find that the T1 fuzzy classifiers are quite similar to the Bayesian classifiers mathematically except the normalization factor $\frac{1}{\sqrt{(2\pi)^d |\mathbf{\Sigma}|}}$ in Eq. (3.63). Usually, such a normalization factor does not affect the classification results so that no essential difference exists between T1 fuzzy classifiers and Bayesian classifiers. However, by introducing data and model uncertainties, the T2 fuzzy classifiers make a different decision Eq. (3.60) from that of the Bayesian classifiers. This is another reason why we prefer T2 FSs to T1 FSs for pattern recognition problems.

## 3.5 The Type-2 Fuzzy Bayesian Decision Theory

In the labeling problem, the linguistic labels for human perceptions are fuzzy because labels may mean different things to different people. As shown in Fig. 2.1, some people may assign the label 4 to the observation $o_5$, and some may assign the label 5. The T2 FS is able to directly evaluate the fuzziness of labels by the secondary MF and FOU. As shown in Fig. 3.9a and b, different people may think that the same label has different distributions (primary MFs) denoted by the solid line and dotted line. Such fuzziness of labels may be described by the union of all the distributions with different possibilities (secondary MFs) as shown in Fig. 3.9c and d, where the shaded region is the FOU. As far as the generalization ability is concerned, Fig. 3.9 also illustrates the importance of modeling uncertainty to achieve a better generalization to unknown test data.

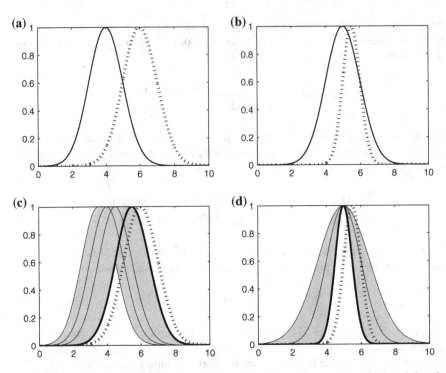

**Fig. 3.9** In **a** and **b**, the distribution of the training data and test data are the *solid line* and *dotted line*. Because of incomplete information and noise, the two distributions are not close. In **c** and **d**, by introducing uncertainty in the model, i.e., letting the model move in a certain way, one of the models (the *thick solid line*) is probably to approximate the test data distribution. The shaded region is the "footprint" of the model uncertainty

Section 2.2 formulates how to search the best labeling configuration based on the Bayesian decision theory. In statistical pattern recognition, we often use probability density functions (PDFs) as likelihood functions to characterize *randomness* of observations. Such uncertainty as randomness can be best encoded in PDFs parameters by the generalized *expectation-maximization* (EM) algorithm from sufficient training data based on the ML criterion (2.7). If randomness is the only uncertainty and can be represented accurately by the model $\lambda$, the maximum posterior probability $P(\mathscr{S}^*|\mathbf{O}, \lambda)$ is the best decision point with the lowest probability of classification error, namely the Bayesian decision theory in Eq. (2.10).

However, we often encounter uncertainties that cannot be characterized by randomness as mentioned in Sects. 1.2 and 3.5: (1) Insufficient or noisy training data make parameters of the model $\lambda$ uncertain, so that the mapping of the model $\lambda$ is also uncertain. (2) The relationship between training data and unknown test data is uncertain due to limited prior information. (3) The non-stationarity cannot be described mathematically. (4) The linguistic labels are uncertain since the same observation may mean different things to different people. Such uncertainties may be considered as *fuzziness* resulting from incomplete information, i.e., fuzzy observations (we are not sure if data are sufficient and clean), fuzzy models (we are not sure if parameters imply the correct mapping), and fuzzy labels (different people may interpret the same thing by different labels). Such uncertainty as fuzziness may make PDFs not so good to directly model large variations of patterns due to limited training data and prior information.

The T2 MF can simultaneously evaluate randomness and fuzziness. Because of fuzzy data, fuzzy models, and fuzzy labels, distributions (primary MFs) are fuzzy denoted either by the solid line or by the dotted line as shown in Fig. 3.9a and b. We may unite all uncertain distributions with different possibilities (secondary MFs) to account for this fuzziness by the bounded FOU as shown in Fig. 3.9c and d. Due to the three-dimensional T2 MF, we use the primary MF to represent the likelihood of fuzzy labels as well as prior distributions, and use the secondary MF to evaluate the fuzziness of this primary MF. Hence, the third dimension of the T2 MF provides additional degrees of freedom to handle fuzziness of distributions characterized by randomness within a unified framework. After specifying the bounds of the FOU and the corresponding secondary MFs, both model and data uncertainties propagate in the system through T2 FS operations, and their effects are evaluated by the type-reduced and defuzzified value of the output T2 FS. This strategy of modeling uncertainty has found its utility in a broad spectrum of applications of interest to signal processing, especially when the signal is non-stationary and cannot be expressed ahead of time mathematically.

A fuzzy labeling problem is specified in terms of a set of sites, $1 \leq i \leq I$, and a set of linguistic labels, $1 \leq j \leq J$; the $j$th label at any site $i$ is denoted by $\tilde{s}_i = j$. The tilde means that we are not sure if the $j$th label is assigned to the $i$th site. A particular fuzzy labeling configuration at all sites is denoted by $\mathscr{S} = \{\tilde{s}_1, \tilde{s}_2, \ldots, \tilde{s}_I\}$. Based on the Bayesian decision theory, we extend the MAP criterion from Eqs. (2.10) and (2.12) using T2 FSs operations,

$$\mathscr{S}^* = \arg\max_{\mathscr{S}} h_{\tilde{\lambda}}(\mathscr{S}|\mathbf{O}), \tag{3.67}$$

$$h_{\tilde{\lambda}}(\mathscr{S}|\mathbf{O}) \propto h_{\tilde{\lambda}}(\mathbf{O}|\mathscr{S}) \sqcap h_{\tilde{\lambda}}(\mathscr{S}), \tag{3.68}$$

where $\tilde{\lambda}$ is the fuzzy model due to incomplete information, $\sqcap$ is the meet operator, $\mathscr{S} = \{\tilde{s}_1, \ldots, \tilde{s}_I\}$ is any fuzzy labeling configuration, $h_{\tilde{\lambda}}(\mathbf{O}|\mathscr{S})$ is the T2 MF of observations $\mathbf{O}$ to $\mathscr{S}$, $h_{\tilde{\lambda}}(\mathscr{S})$ is the T2 MF evaluating our belief to $\mathscr{S}$, and $h_{\tilde{\lambda}}(\mathscr{S}|\mathbf{O})$ is our belief to $\mathscr{S}$ given observations $\mathbf{O}$. The maximum $h_{\tilde{\lambda}}(\mathscr{S}|\mathbf{O})$ may be defined according to different type-reduction and defuzzification methods, such as centroids of T2 FSs in Sect. 3.3. Moreover, the nonsingleton fuzzification is able to handle fuzzy data due to noise as mentioned in Sect. 3.1. Similar to Eq. (2.14), we continue to assume that the observations are independent, so $h_{\tilde{\lambda}}(\mathbf{O}|\mathscr{S})$ is written as a meet of T2 MFs of individual observations,

$$h_{\tilde{\lambda}}(\mathbf{O}|\mathscr{S}) = \sqcap_{i=1}^{I} h_{\tilde{\lambda}}(o_i|\tilde{s}_i = j), 1 \leq j \leq J. \tag{3.69}$$

The major reason why we use T2 FSs rather than T1 FSs to handle uncertainties lies in three aspects: (1) T2 FSs represent more uncertainties simultaneously by primary and secondary MFs; (2) T2 FSs handle bounded uncertainty (FOU) by T2 FSs operations more effectively than T1 FSs operations; and (3) different type-reduction and defuzzification techniques of T2 FSs may produce different results, which provides additional flexibility to solve pattern recognition problems.

For simplicity, we use IT2 MFs with uniform possibilities over the entire FOU. The IT2 MF is represented by interval sets of lower MFs and upper MFs denoted by $h_{\tilde{\lambda}}(\cdot) = [\underline{h}_{\tilde{\lambda}}(\cdot), \overline{h}_{\tilde{\lambda}}(\cdot)]$. Therefore, from Eq. (3.68), all uncertainties in $h_{\tilde{\lambda}}(\mathbf{O}|\tilde{\mathscr{F}})$ and $h_{\tilde{\lambda}}(\tilde{\mathscr{F}})$, like fuzzy data, fuzzy models, and fuzzy labels, are translated into the IT2 MF $h_{\tilde{\lambda}}(\tilde{\mathscr{F}}|\mathbf{O}) = [\underline{h}_{\tilde{\lambda}}(\tilde{\mathscr{F}}|\mathbf{O}), \overline{h}_{\tilde{\lambda}}(\tilde{\mathscr{F}}|\mathbf{O})]$. Hence the original decision boundary by the Bayesian decision theory is extended to a decision interval set, which may include more information to lower the classification error when both fuzziness and randomness exist.

To make the best decision based on the interval set $h_{\tilde{\lambda}}(\tilde{\mathscr{F}}|\mathbf{O})$ is a central issue of T2 FSs for handling uncertainty in pattern recognition. Without loss of generality, let us consider a two-category classification problem as shown in Fig. 3.10. Solid lines denote the posterior distributions $P(\tilde{\lambda}|\mathbf{O})$ of classes $\tilde{\lambda}_1$ and $\tilde{\lambda}_2$. According to the Bayesian decision theory, the intersection point $b$ of two lines is the best decision boundary for the minimum probable error rate,

$$\text{Decide } \lambda_1 \text{ if } P(\lambda_1|\mathbf{O}) > P(\lambda_2|\mathbf{O}). \tag{3.70}$$

As shown in Fig. 3.10, posterior distributions are uncertain within the FOU bounded by lower and upper MFs because of fuzziness. To minimize the overall error rate, we obtain the decision rule,

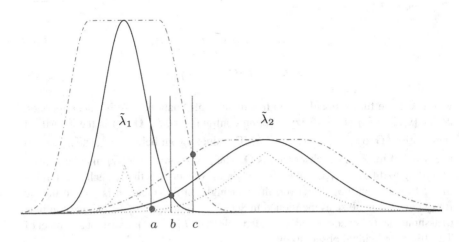

**Fig. 3.10** Type-2 fuzzy Bayesian decision theory

$$\text{Decide } \tilde{\lambda}_1 \text{ if } h_{\tilde{\lambda}_1}(\mathbf{O}) > h_{\tilde{\lambda}_2}(\mathbf{O}). \tag{3.71}$$

This rule is represented by lower MFs and upper MFs as follows,

$$\text{Decide } \tilde{\lambda}_1 \text{ if } \underline{h}_{\tilde{\lambda}_1}(\mathbf{O}) > \underline{h}_{\tilde{\lambda}_2}(\mathbf{O}) \text{ and } \overline{h}_{\tilde{\lambda}_1}(\mathbf{O}) > \overline{h}_{\tilde{\lambda}_2}(\mathbf{O}). \tag{3.72}$$

As shown in Fig. 3.10, lower MFs (dotted lines) of $\tilde{\lambda}_1$ and $\tilde{\lambda}_2$ intersect at point $a$, and upper MFs (dashed lines) of $\tilde{\lambda}_1$ and $\tilde{\lambda}_2$ intersect at point $c$. From the decision rule Eq. (3.72), all samples on the left of the point $a$ is classified into $\tilde{\lambda}_1$, and all samples on the right of the point $c$ is classified into $\tilde{\lambda}_2$. However, samples between $a$ and $c$ are difficult to classify, because if we classify them based on lower MFs, the error rate decided by upper MFs will increase; and on the contrary, if we classify them based on upper MFs, we simultaneously increase the error rate by lower MFs. Therefore, between $a$ and $c$, no decision boundaries can reduce the error induced by both lower and upper MFs. Often samples between $a$ and $c$ are similar patterns that are difficult to distinguish. Many misclassifications result from these samples.

To achieve the minimum classification error, we may select the decision boundary that minimizes the sum of errors induced by both lower and upper MFs. However, this solution has two practical problems: (1) Many pattern recognition problems have complicated fuzzy posterior distributions in high-dimensional feature space, so it is almost impossible to find the exact decision boundary analytically. (2) Because lower MFs are always smaller than corresponding upper MFs, the error induced by lower MFs is smaller than upper MFs, which results in a decision boundary much closer to the one decided by upper MFs. Therefore, we have to balance the error caused by lower and upper MFs especially for similar patterns.

Because T2 FSs incorporate reasonable prior knowledge about fuzziness in pattern recognition, they will provide additional decision information when similar patterns are hard to distinguish. The decision boundary is fuzzy because different people may classify patterns by different decision boundaries instead of the maximum posterior. Such fuzziness happens especially when people classify similar patterns. Obviously, the fuzzy label $\tilde{s}_i$ includes labels decided by lower MFs, centroids, and upper MFs, $\tilde{s}_i = \{\underline{s}_i, s_i, \overline{s}_i\}$, which reflects that people may interpret the same thing by different labels.

## 3.6 Summary

The T2 FS was introduced initially by Zadeh as an extension of the T1 FS [1]. A thorough investigation of properties of T2 FSs and higher types was done by Mizumoto and Tanaka [2, 3]. Klir and Folger discussed that the T1 MFs seem problematical, because a representation of fuzziness is made using membership grades that are themselves precise real numbers [4]. Thus, they extended the concept of T1 FSs to T2 FSs and further to higher types of FSs [4]. Especially, they called IT2

FSs as interval-valued FSs [4]. Mendel and John introduced the new terminology to distinguish between the T1 and T2 MFs, by which they represented the T2 MF in vertical-slice and wavy-slice manners respectively [5]. Based on the wavy-slice representation, they also proposed the concept of embedded FSs [5], which justifies the modeling capability of T2 FSs and FLSs for handling uncertainty [6]. Mitchell ranked the T2 fuzzy numbers by ranking the embedded T1 fuzzy numbers associated with different weights [7]. To learn more about T2 FSs quickly (including the history of T2 FSs), Mendel's "Type-2 Fuzzy Sets: Some Questions and Answers" is worth reading [8]. Mendel also summarized the current (until year 2001) development and application of T2 FSs and FLSs in his book [9].

Set operations are foundations in the theory of T2 FSs. Mizumoto and Tanaka studied the set theoretic operations and properties of T2 FSs [2]. Karnik and Mendel extended the works of Mizumoto and Tanaka, and obtained practical algorithms for performing union, intersection, and complement for T2 FSs [10, 11]. Mendel and John reformulated all set operations in both vertical-slice and wavy-slice manners [5]. They concluded that the general T2 FSs operations are too complex to implement in practice, but the IT2 FSs operations are simple so that they are the most widely used T2 FSs [5].

Based on T2 FSs, uncertain rule-based FLSs present an expanded and richer fuzzy logic for real-world problems [9]. Karnik, Mendel, and Liang established a complete T2 FLS theory [10, 12–16]. Liang and Mendel developed a complete theory for the IT2 FLS [17]. The T2 FLS includes five parts: fuzzifier, rule base, inference engine, type reducer, and defuzzifier. Mousouris and Mendel explored the nonsingleton fuzzifier [18]. The rule base, inference engine, and type reducer have been studied [15], where the concept of the centroid of a T2 FS has been developed in the type-reduction. Inspired by the corresponding T1 FLS defuzzifier and the concept of embedded sets, Karnik and Mendel presented the centroid, hight, and center-of-sets type reducers, and provided a practical algorithm (the KM algorithm) for computing type-reduced set (centroid) for IT2 FSs [10, 16, 17]. Wu and Mendel interpreted the KM algorithm by information theory, and demonstrated that the centroid of the IT FS can measure the uncertainty of the output of an IT2 FLS. Furthermore, Mendel and Liu discovered the fast convergence of the KM algorithm, and proved it in the case of the continuous representation of IT2 FSs [19]. The IT2 FLS have been receiving increasing attention recently [20, 21].

The recent success of T2 FSs in control, signal processing, and pattern recognition has largely been attributed to its three-dimensional MFs to handle uncertainties existing universally in video streams [22], time series [17, 18], communication channels [15, 23, 24], control of mobile robots [25], linguistic meaning of words [26], human perceptions [27], and patterns like speech, handwriting, and face [28–33]. The T2 MF represents the uncertainties by two fundamental concepts: *secondary MF* and *FOU* [5]. Having the domain in the interval [0, 1], the secondary MF is a function of the membership (not just a point of value as type-1 (T1) MF) at each value of the input primary variable [10]. The union of all the secondary MF domain composes a bounded region FOU reflecting the degree of uncertainty of the model (the shaded region in Fig. 3.1) [5]. Uncertainties exist in both *data* and *model* [34]. After specifying the uncertain bounds of the FOU and the corresponding secondary MFs, both model and data uncertainties propagate through T2 FSs operations, and their effects are evaluated using the type-reduced set and defuzzified value of the output T2 FS [9]. IT2 MFs have been widely used because of their computational efficiency [5, 17]. Liang and Mendel proposed two types of IT2 MFs based on Gaussian primary MFs [17]. Zeng and Liu extended their work to multivariate Gaussian primary MFs for high-dimensional feature vectors [33].

Mendel mentioned that T2 FSs may be applicable when [5, 8]:

- The data-generating system is known to be time-varying, but the mathematical description of the time-variability is unknown (e.g., as in mobile communications);
- Measurement noise is non-stationary, and the mathematical description of the non-stationarity is unknown (e.g., as in a time-varying SNR);
- Features in a pattern recognition application have statistical attributes that are non-stationary, and the mathematical descriptions of the non-stationarity are unknown;
- Knowledge is mined from a group of experts using questionnaires that involve uncertain words;
- Linguistic terms are used that have a nonmeasurable domain.

A lot of applications of T2 FSs and FLSs solve one or more of above points in the real-world problems. The official web site for T2 FSs and FLSs lists their most recent developments and applications [35]. The T2 FSs and FLSs may challenge us in two aspects. Mathematically, how do we justify the validity of using three-dimensional MFs, in which one dimension describes the uncertainty of another? Practically, how do we design the secondary MFs and FOU according to the problems at hand, and reduce the computational complexity of the T2 FSs operations in the meanwhile? The answers will be the future research directions of T2 FSs and FLSs.

The fuzzy classifier design can be found in [36]. Pattern recognition using T2 FSs and FLSs has been investigated in the past decade [22, 28–33, 37–41]. The basic idea is to describe both data (using the T1 or T2 nonsingleton fuzzification) and model (using the secondary MF and FOU) uncertainties within a unified framework. Mitchell viewed the task of pattern recognition as the similarity measure between two T2 FSs, one for the uncertainty in the feature space, the other for the uncertainty in the hypothesis space [28]. Liang and Mendel applied the IT2 FLS to video traffic classification (two classes), where they used the center of centroid of the output IT2 FS to make decision [22]. They compared three classifiers, namely the Bayesian classifier, the T1 fuzzy classifier, and the T2 fuzzy classifier. In terms of mathematical expressions, we find that the Bayesian classifier is quite similar to the T1 fuzzy classifier, while T2 fuzzy classifier adds additional uncertainty description using the third dimension of the T2 MF. Non-stationarities with unknown mathematical description is a kind of uncertainty. From [22], we find another reason for using T2 FSs: The features have statistical attributes that are non-stationary, which can be handled by the T2 FS more effectively. Zeng and Liu integrated T2 FSs with Markov processes for speech and handwriting recognition [32]. They used one dimension (primary MF) of the T2 MF to model the randomness, and used another dimension (secondary MF) to describe the fuzziness of the primary MF. Obviously, randomness and fuzziness are two important uncertainties in pattern recognition problems. In conclusion, we have till now three reasons to justify the T2 FS and FLS for pattern recognition all from the "uncertainty" perspective. The relationships among fuzzy sets, uncertainty, and information have been fully discussed in [4].

In the classical book [4], Klir and Folger state that one way of simplifying a very complex system is to allow some degree of uncertainty in its description. Regarding the *nature of uncertainty*, they state: "Three types of uncertainty are now recognized ... *fuzziness* (or vagueness), which results from the imprecise boundaries of FSs; *non-specificity* (or information-based imprecision), which is connected with sizes (cardinalities) of relevant sets or alternatives; and *strife* (or discord), which expresses conflicts among the various sets of alternatives." The theory of fuzzy sets and fuzzy measures can properly characterize and investigate these three types of uncertainties.

We shall distinguish between two high-level kinds of uncertainties, *random* and *fuzzy*. Probability theory is associated with the former, and FSs can be associated with the latter. If FSs are used in applications in which randomness is present, then *both uncertainties should be accounted for* [8]. Ross discussed the fuzzy theory and probability theory especially from the Bayesian view [42]. Liu established uncertainty theory that handles random and fuzzy uncertainties simultaneously in the system [43, 44]. He also proposed the concepts of random fuzzy variable and random fuzzy variable [44]. Möller and Beer applied *fuzzy randomness* to civil engineering [34]. Buckley combined fuzzy sets and probability leading to the fuzzy probability and fuzzy statistics [45, 46]. In his book [45], he also proposed the fuzzy Markov chains with fuzzy transition probability. In this chapter, we present T2 FSs for handling fuzziness and randomness in pattern recognition problems [32]. Because of fuzziness, the classical Bayesian decision theory is no longer the best decision rule for classification. Thus, we extend it by T2 FS operations, in which the fuzzy likelihood and prior are naturally connected by "$\sqcap$" operation.

To enhance the expressive power for uncertainty, we shall investigate four type-2 fuzzy graphical models, type-2 Gaussian mixture models (T2 FGMMs) [47], type-2 fuzzy hidden Markov models (T2 FHMMs) [29, 30, 33], type-2 fuzzy Markov random fields (T2 FMRFs) [31, 48], and type-2 fuzzy latent Dirichlet allocation (T2 FLDA) [49]. The T2 FHMM is suitable for uncertain sequential data modeling, while the T2 FHMR is good at modeling two-dimensional uncertain relationships of image regions. In the following chapters, we shall apply type-2 fuzzy graphical models to many pattern recognition problems, such as speech, handwriting, and human motion recognition.

# References

1. Zadeh, L.A.: The concept of a linguistic variable and its application to approximate reasoning. Inf. Sci. **8**, 199–249 (1975)
2. Mizumoto, M., Tanaka, K.: Some properties of fuzzy sets of type-2. Inf. Control **31**, 312–340 (1976)

3. Mizumoto, M., Tanaka, K.: Fuzzy sets of type-2 under algebraic product and algebraic sum. Fuzzy Sets Syst. **5**, 277–290 (1981)
4. Klir, G.J., Folger, T.A.: Fuzzy Sets, Uncertainty, and Information. Prentice-Hall, Englewood Cliffs (1988)
5. Mendel, J.M., John, R.I.B.: Type-2 fuzzy sets made simple. IEEE Trans. Fuzzy Syst. **10**(2), 117–127 (2002)
6. Wu, D., Tan, W.: Type-2 FLS modeling capability analysis. In: Proceedings of the FUZZ-IEEE, pp. 242–247 (2005)
7. Mitchell, H.: Ranking type-2 fuzzy numbers. IEEE Trans. Fuzzy Syst. **14**(2), 287–294 (2006)
8. Mendel, J.M.: Type-2 fuzzy sets: Some questions and answers. IEEE Connections, Newsl. IEEE Neural Netw. Soc. **1**, 10–13 (2003)
9. Mendel, J.M.: Uncertain Rule-based Fuzzy Logic Systems: Introduction and New Directions. Prentice-Hall, Upper Saddle River (2001)
10. Karnik, N.N., Mendel, J.M.: An introduction to type-2 fuzzy logic systems (1998). http://sipi. usc.edu/~mendel/report
11. Karnik, N.N., Mendel, J.M.: Operations on type-2 fuzzy sets. Fuzzy Sets Syst. **122**, 327–348 (2001)
12. Karnik, N.N., Mendel, J.M.: Introduction to type-2 fuzzy logic systems. In: Proceedings of the FUZZ-IEEE, pp. 915–920 (1998)
13. Karnik, N.N., Mendel, J.M.: Type-2 fuzzy logic systems: Type-reduction. In: Proceedings of the IEEE SMC, pp. 2046–2051 (1998)
14. Karnik, N.N.: Type-2 fuzzy logic systems. Ph.D. thesis, University of Southern California, Los Angeles (1998)
15. Karnik, N.N., Mendel, J.M., Liang, Q.: Type-2 fuzzy logic systems. IEEE Trans. Fuzzy Syst. **7**(6), 643–658 (1999)
16. Karnik, N.N., Mendel, J.M.: Centroid of a type-2 fuzzy set. Inf. Sci. **132**, 195–220 (2001)
17. Liang, Q., Mendel, J.M.: Interval type-2 fuzzy logic systems: Theory and design. IEEE Trans. Fuzzy Syst. **8**(5), 535–549 (2000)
18. Mouzouris, G.C., Mendel, J.M.: Nonsingleton fuzzy logic systems: Theory and application. IEEE Trans. Fuzzy Syst. **5**(1), 56–71 (1997)
19. Mendel, J.M., Liu, F.: Super-exponential convergence of the Karnik-Mendel algorithms for computing the centroid of an interval type-2 fuzzy set. IEEE Trans. Fuzzy Syst. p. accepted for publication (2005)
20. Mendel, J.M., John, R.I.B., Liu, F.: Interval type-2 fuzzy logic systems made simple. IEEE Trans. Fuzzy Syst. p. accepted for publication (2005).
21. Mendel, J.M.: Computing derivatives in interval type-2 fuzzy logic systems. IEEE Trans. Fuzzy Syst. **12**(1), 84–98 (2004)
22. Liang, Q., Mendel, J.M.: MPEG VBR video traffic modeling and classification using fuzzy technique. IEEE Trans. Fuzzy Syst. **9**(1), 183–193 (2001)
23. Liang, Q., Mendel, J.M.: Equalization of nonlinear time-varying channels using type-2 fuzzy adaptive filters. IEEE Trans. Fuzzy Syst. **8**(5), 551–563 (2000)
24. Liang, Q., Mendel, J.M.: Overcoming time-varying co-channel interference using type-2 fuzzy adaptive filters. IEEE Trans. Circuits Syst. II **47**(12), 1419–1429 (2000)
25. Hagras, H.: A hierarchical type-2 fuzzy logic control architecture for autonomous mobile robots. IEEE Trans. Fuzzy Syst. **12**(4), 524–539 (2004)
26. Mendel, J.M.: Fuzzy sets for words: a new beginning. In: Proceedings of the FUZZ-IEEE **1**, 37–42 (2003)
27. John, R.I.B.: Perception modeling using type-2 fuzzy sets. Ph.D. thesis, De Montfort University, Leicester (2000)
28. Mitchell, H.: Pattern recognition using type-II fuzzy sets. Inf. Sci. **170**, 409–418 (2005)
29. Zeng, J., Liu, Z.Q.: Interval type-2 fuzzy hidden Markov models. In: Proceedings of the FUZZ-IEEE, pp. 1123–1128 (2004)
30. Zeng, J., Liu, Z.Q.: Type-2 fuzzy hidden Markov models to phoneme recognition. In: 17th International Conference on Pattern Recognition **1**, pp. 192–195 (2004)

31. Zeng, J., Liu, Z.Q.: Type-2 fuzzy Markov random fields to handwritten character recognition. In: 18th International Conference on Pattern Recognition, pp. 1162–1165 (2006)
32. Zeng, J., Liu, Z.Q.: Type-2 fuzzy sets for handling uncertainty in pattern recognition. In: Proceedings of the FUZZ-IEEE, pp. 6597–6602 (2006)
33. Zeng, J., Liu, Z.Q.: Type-2 fuzzy hidden Markov models and their application to speech recognition. IEEE Trans. Fuzzy Syst. **14**(3), 454–467 (2006)
34. Möller, B., Beer, M.: Fuzzy Randomness: Uncertainty in Civil Engineering and Computational Mechanics. Springer, New York (2004)
35. type2fuzzylogic.org. http://www.type2fuzzylogic.org/
36. Kuncheva, L.I.: Fuzzy Classifier Design. Physica-Verlag, Heidelberg (2000)
37. Rhee, F., Hwang, C.: A type-2 fuzzy c-means clustering algorithm. In: Proceedings of the Joint 9th IFSA World Congress and 20th NAFIPS International Conference, pp. 1926–1929 (2001)
38. Wu, H., Mendel, J.: Multi-category classification of ground vehicles based on the acoustic data of multiple terrains using fuzzy logic rule-based classifiers. In: Proceedings of the SPIE Defense and Security Conference, Unattended Ground Sensor Technologies and Applications VII, pp. 52–57 (2005)
39. Wu, H., Mendel, J.: Multi-category classification of ground vehicles using fuzzy logic rule-based classifiers: Early results. In: Proceedings of the 7th IASTED International Conference Artificial Intelligence and Soft Computing, pp. 52–57 (2003)
40. Wu, H., Mendel, J.: Classification of ground vehicles from acoustic data using fuzzy logic rule-based classifiers: Early results. Proc. SPIE Int. Soc. Opt. Eng. **4743**, 62–72 (2002)
41. Wu, H., Mendel, J.: Classifier designs for binary classifications of ground vehicles. Proc. SPIE Int. Soc. Opt. Eng. **5090**(1), 122–133 (2003)
42. Ross, T.J., Booker, J.M., Parkinson, W.J.: Fuzzy Logic and Probability Applications: Bridging the Gap, 2nd edn. Society for Industrial and Applied Mathematics, Philadelphia (2002)
43. Liu, B.: Theory and Practice of Uncertain Programming. Physica-Verlag, New York (2002)
44. Liu, B.: Uncertainty Theory: An Introduction to Its Axiomatic Foundations. Springer, Berlin (2004)
45. Buckley, J.J.: Fuzzy Probabilities: New Approach and Applications. Physica-Verlag, New York (2003)
46. Buckley, J.J.: Fuzzy Statistics. Springer, New York (2004)
47. Zeng, J., Xie, L., Liu, Z.Q.: Type-2 fuzzy Gaussian mixture models. Pattern Recogn. **41**(12), 3636–3643 (2008)
48. Zeng, J., Liu, Z.Q.: Type-2 fuzzy Markov random fields and their application to handwritten Chinese character recognition. IEEE Trans. Fuzzy Syst. **16**(3), 747–760 (2008)
49. Cao, X.Q., Liu, Z.Q.: Type-2 fuzzy topic models for human action recognition. IEEE Trans. Fuzzy Syst. (2013, submitted)

# Chapter 4
# Type-2 Fuzzy Gaussian Mixture Models

**Abstract** This chapter presents a new extension of Gaussian mixture models (GMMs) based on type-2 fuzzy sets (T2 FSs) referred to as T2 FGMMs. The estimated parameters of the GMM may not accurately reflect the underlying distributions of the observations because of insufficient and noisy data in real-world problems. By three-dimensional membership functions of T2 FSs, T2 FGMMs use footprint of uncertainty (FOU) as well as interval secondary membership functions to handle GMMs uncertain mean vector or uncertain covariance matrix, and thus GMMs parameters vary anywhere in an interval with uniform possibilities. As a result, the likelihood of the T2 FGMM becomes an interval rather than a precise real number to account for GMMs uncertainty. These interval likelihoods are then processed by the generalized linear model (GLM) for classification decision-making. In this chapter, we focus on the role of the FOU in pattern classification. Multi-category classification on different datasets from UCI repository shows that T2 FGMMs are consistently as good as or better than GMMs in case of insufficient training data, and are also insensitive to different areas of the FOU.

## 4.1 Gaussian Mixture Models

Gaussian mixture models (GMMs) have been widely used in density modeling and clustering. They have universal approximation ability because they can model any density function closely provided that they contain enough mixture components [1]. The multivariate Gaussian distribution is

$$N(\mathbf{o}; \mu, \Sigma) = \frac{1}{\sqrt{(2\pi)^d |\Sigma|}} e^{-\frac{1}{2}(\mathbf{o}-\mu)^{\mathrm{T}} \Sigma^{-1}(\mathbf{o}-\mu)}, \tag{4.1}$$

where $\mu$ is the mean vector, $\Sigma$ is the covariance matrix, and $d$ is the dimensionality of $\mathbf{o}$. In this chapter, we study only the diagonal covariance matrix, $\Sigma = \mathrm{diag}(\sigma_1^2, \ldots, \sigma_d^2)$. As far as the non-diagonal symmetric $\Sigma$ is concerned, we have the orthogonal matrix $Q$ and diagonal matrix $\Lambda$, so that $\Sigma^{-1} = Q\Lambda^{-1}Q^{\mathrm{T}}$.

© Tsinghua University Press, Beijing and Springer-Verlag Berlin Heidelberg 2015
J. Zeng and Z.-Q. Liu, *Type-2 Fuzzy Graphical Models for Pattern Recognition*,
Studies in Computational Intelligence 591, DOI 10.1007/978-3-662-44690-4_4

In this way, we can still deal with diagonal matrix by transforming the observation vector to $\mathbf{o}Q^\mathrm{T}$ and mean vector to $\mu Q^\mathrm{T}$. The GMM is composed of $M$ mixture components of multivariate Gaussian as follows,

$$p(\mathbf{o}) = \sum_{m=1}^{M} w_m N(\mathbf{o}; \mu_m, \Sigma_m), \tag{4.2}$$

where the mixing weight, $\sum_{m=1}^{M} w_m = 1, w_m > 0$.

In GMM pattern classification, we would have a number of GMMs, $1 \leq j \leq J$, one for each category, and classify a test observation $\mathbf{o}$ according to the model with the highest class-conditional probability $p(\mathbf{o}|j)$, which is called the likelihood of the $j$th GMM with respect to $\mathbf{o}$. The expectation-maximization (EM) algorithm [2] estimates GMMs parameters from a dataset according to the maximum-likelihood (ML) criterion. The GMM is completely certain once its parameters are specified. However, because of insufficient or noisy data in real-world problems, the GMM may not accurately reflect the underlying distribution of the observations according to the ML estimation. It may seem problematical to use likelihoods that are themselves precise real numbers to evaluate GMMs with uncertain parameters. Although this does not pose a serious problem for many applications, it is nevertheless possible to describe GMMs uncertain parameters for uncertain likelihoods. Therefore, introducing descriptions of uncertain parameters should allow us to make robust decisions in pattern classification.

Type-2 fuzzy sets (T2 FSs) [3–6] provide a theoretically well-founded framework to handle GMMs uncertain parameters. Their recent success achieved in pattern recognition has been largely attributed to their three-dimensional membership functions (MFs) for modeling uncertainties [7, 8] (e.g., radiographic tibia image data clustering [9], MPEG VBR video traffic modeling and classification [10], automatic welded structure classification [11], image thresholding [12], speech recognition [7], battle ground vehicle classification [13], and handwritten Chinese character recognition [14]). Within the T2 FS framework, T2 MFs use one dimension, the primary membership, to describe the observation uncertainty, and use another dimension, the secondary MF, to evaluate the uncertainty of that primary membership. Figure 4.1 shows a typical T2 MF—the Gaussian primary MF with uncertain mean, $\mu \in [\underline{\mu}, \overline{\mu}]$, which varies within an interval having uniform possibilities in Fig. 4.1a. The observation $o$'s primary membership is an interval bounded by lower and upper membership grades (See Fig. 4.1b, $[\underline{h}(o), \overline{h}(o)]$, whose uncertainty is further described by the secondary MF in Fig. 4.1c. The primary membership also has uniform possibilities leading to the interval secondary MF because the mean varies with uniform possibilities. The union of all primary memberships composes the footprint of uncertainty (FOU) shown as the shaded region in Fig. 4.1b. The operations on T2 FSs [15] can propagate the uncertainty covered by the FOU for final decisions. More specifically, in case of interval secondary MFs, T2 FS operations involve only simple interval computations [16]. For example, if two interval primary membership grades are $[\underline{h}_1, \overline{h}_1]$

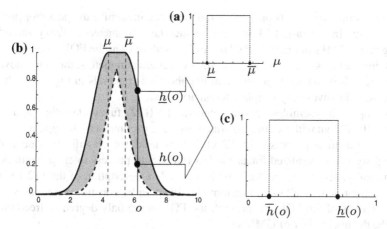

**Fig. 4.1** The Gaussian primary membership function with uncertain mean in **a**, which has *lower* (*thick dashed line*) and *upper* (*thick solid line*) boundaries in **b**. The primary membership is an interval, $[\underline{h}(o), \overline{h}(o)]$, with uniform possibilities (interval secondary membership function) in **c**. The *shaded region* in **b** is the FOU

and $[\underline{h}_2, \overline{h}_2]$, then their sum and product are $[\underline{h}_1 + \underline{h}_2, \overline{h}_1 + \overline{h}_2]$ and $[\underline{h}_1\underline{h}_2, \overline{h}_1\overline{h}_2]$, respectively.

The GMM is parameterized by two moments—the mean vector $\mu$ and covariance matrix $\Sigma$, whose fuzziness can be described by T2 FSs referred to as type-2 fuzzy GMMs (T2 FGMMs). To this end, we assume the mean or the variance vary within an interval with uniform possibilities, which can be represented by the T2 MF in Fig. 4.1. Therefore, the likelihood of the T2 FGMM becomes an interval with uniform possibilities in Fig. 4.1c, which is composed of every likelihood of the embedded GMM in the FOU with uniform weighting.

The T2 FGMM is different from the fuzzy GMM (FGMM) [17] that estimates its parameters based on the modified fuzzy c-means algorithm [18]. Rather than modeling GMMs uncertain parameters, the FGMM focuses on the precise parameter estimation of GMMs using fuzzy approaches [19, 20]. The Generalized Gaussian (GG) distribution introduces one additional shape parameter compared to the Gaussian model, and it can approximate a large class of statistical distributions [21]. Indeed, the shape parameter tunes the decay rate of the density function, and thus describes the shape variation of the distribution. However, the GG distribution has the precise parameters after training, unlike the fuzzy parameters and the FOU introduced in the T2 FGMM. Furthermore, the T2 FGMM uses interval likelihoods to describe the observation uncertainty, which is different from the shape parameter in the GG distribution. Contrary to the T2 fuzzy logic system (FLS) [22] using GMMs as secondary membership functions to generate secondary membership grades, the proposed T2 FGMMs consider GMMs as the primary membership and use the interval secondary grades to describe the uncertainty of GMMs. At present, Bayesian methods [23] view uncertain parameters as random variables having some known distribution, and training data convert this distribution on the variables into posterior probability density.

Bayesian estimation selects only the parameters that maximize the posterior probability density. In contrast, T2 FSs describe uncertain parameters as fuzzy variables having interval MFs in Fig. 4.1a, and unite all possibilities as the FOU to account for uncertainty in Fig. 4.1b. In modeling GMMs uncertain parameters, the major advantage of T2 FSs is to represent uncertain likelihoods as intervals in Fig. 4.1c, whose computation involves only simple interval arithmetic.

Our approach resembles previous interval T2 (IT2) fuzzy classifiers and systems [7, 10, 13] but with two main distinctions: first, in contrast to the type-reduction and defuzzification processes in T2 classifiers used for classification decision-making, we use generalized linear models (GLMs) to process interval likelihoods automatically; second, since the FOU is a complete description of the IT2 FS, we focus on the role of the FOU in pattern classification by constructing models with different areas of the FOU. At present, the FOU is the only degree of freedom to describe the uncertainty of GMMs.

## 4.2  Type-2 Fuzzy Gaussian Mixture Models

The GMM (4.2) is expressed as a linear combination of multivariate Gaussian distributions (4.1). So, we use T2 MFs to represent multivariate Gaussian with uncertain mean vector or covariance matrix, and replace the corresponding parts in (4.2) to produce the T2 FGMM with uncertain mean vector (T2 FGMM-UM) or uncertain variance (T2 FGMM-UV). Given a $d$-dimensional observation vector $\mathbf{o}$, the mean vector $\mu$, and the diagonal covariance matrix $\Sigma = \mathrm{diag}(\sigma_1^2, \ldots, \sigma_d^2)$, the multivariate Gaussian with uncertain mean vector or covariance matrix is

$$N(\mathbf{o}; \tilde{\mu}, \Sigma) = \frac{1}{\sqrt{(2\pi)^d |\Sigma|}} \exp\left[-\frac{1}{2}\left(\frac{o_1 - \mu_1}{\sigma_1}\right)^2\right] \cdots \exp\left[-\frac{1}{2}\left(\frac{o_d - \mu_d}{\sigma_d}\right)^2\right],$$
$$\mu_1 \in [\underline{\mu}_1, \overline{\mu}_1], \ldots, \mu_d \in [\underline{\mu}_d, \overline{\mu}_d], \tag{4.3}$$

or

$$N(\mathbf{o}; \mu, \tilde{\Sigma}) = \frac{1}{\sqrt{(2\pi)^d |\Sigma|}} \exp\left[-\frac{1}{2}\left(\frac{o_1 - \mu_1}{\sigma_1}\right)^2\right] \cdots \exp\left[-\frac{1}{2}\left(\frac{o_d - \mu_d}{\sigma_d}\right)^2\right],$$
$$\sigma_1 \in [\underline{\sigma}_1, \overline{\sigma}_1], \ldots, \sigma_d \in [\underline{\sigma}_d, \overline{\sigma}_d], \tag{4.4}$$

where $\tilde{\mu}$ and $\tilde{\Sigma}$ denote uncertain mean vector and covariance matrix, respectively. Because we have no prior knowledge about the parameter uncertainty, practically we assume that the mean and standard deviation (std) vary within intervals with uniform possibilities, i.e., $\mu \in [\underline{\mu}, \overline{\mu}]$ or $\sigma \in [\underline{\sigma}, \overline{\sigma}]$. Obviously, each exponential component in (4.3) and (4.4) is the Gaussian primary MF with uncertain mean or std in Fig. 4.2a or b [16]. From Fig. 4.2 we see that T2 MFs seem to be an ensemble of infinite embedded type-1 (T1) MFs, which have different mean or std to form the shaded FOU. Similarly, we have infinite embedded Gaussian distributions in (4.3) with different mean vectors. However, for simplicity, we only obtain embedded non-Gaussian distributions with different std values in (4.4) because we use the fixed normalization factor $\sqrt{(2\pi)^d |\Sigma|}$ without considering the uncertain determinant $|\tilde{\Sigma}|$ here.

In the Gaussian primary MF with uncertain mean, the upper MF is

$$\overline{h}(o) = \begin{cases} f(o; \underline{\mu}, \sigma), & o < \underline{\mu}; \\ 1, & \underline{\mu} \leq o \leq \overline{\mu}; \\ f(o; \overline{\mu}, \sigma), & o > \overline{\mu}, \end{cases} \tag{4.5}$$

where

$$f(o; \mu, \sigma) \triangleq \exp\left[-\frac{1}{2}\left(\frac{o - \mu}{\sigma}\right)^2\right]. \tag{4.6}$$

The lower MF is

$$\underline{h}(o) = \begin{cases} f(o; \overline{\mu}, \sigma), & o \leq \frac{\underline{\mu} + \overline{\mu}}{2}; \\ f(o; \underline{\mu}, \sigma), & o > \frac{\underline{\mu} + \overline{\mu}}{2}. \end{cases} \tag{4.7}$$

In the Gaussian primary MF with uncertain std, the upper MF is

$$\underline{h}(o) = f(o; \mu, \overline{\sigma}), \tag{4.8}$$

and the lower MF is

$$\overline{h}(o) = f(o; \mu, \underline{\sigma}). \tag{4.9}$$

The factors $k_m$ and $k_v$ control the intervals in which the parameters vary,

$$\underline{\mu} = \mu - k_m \sigma, \quad \overline{\mu} = \mu + k_m \sigma, \quad k_m \in [0, 3], \tag{4.10}$$

$$\underline{\sigma} = k_v \sigma, \quad \overline{\sigma} = \frac{1}{k_v}\sigma, \quad k_v \in [0.3, 1]. \tag{4.11}$$

Because a one-dimensional gaussian has 99.7 % of its probability mass in the range of $[\mu - 3\sigma, \mu + 3\sigma]$, we constrain $k_m \in [0, 3]$ and $k_v \in [0.3, 1]$. These factors also control the area of the FOU. The bigger (or smaller) the $k_m$ (or $k_v$), the larger the FOU, which implies the greater uncertainty in the IT2 FS. Because we use IT2 FSs in practice, the FOU is the only degree of freedom to account for the uncertainty. In this chapter, we shall study the role of the FOU to handle uncertainties in pattern classification.

After introducing the FOU, the likelihood of the T2 FGMM becomes an interval set rather than a precise real number. In information theory [24], the uncertainty of a random variable is measured by its entropy, which is not suitable to measure the uncertainty of the interval set with uniform possibilities. Similarly to the entropy of a uniform random variable, the uncertainty of the interval set is equal to the logarithm of

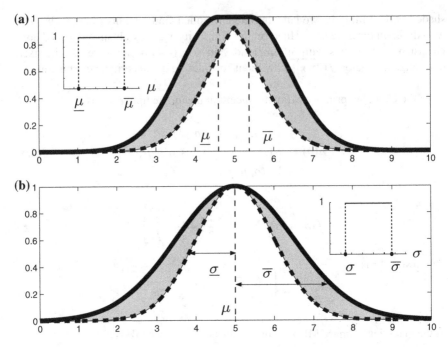

**Fig. 4.2** The Gaussian primary MF with **a** uncertain mean and **b** uncertain std having uniform possibilities. The *shaded region* is the FOU. The *thick solid* and *dashed lines* denote the *lower* and *upper* MFs

the length of the interval [25]. So, the uncertainty of the interval set is logarithmically proportional to its length. For analytical purpose, we often use the *log-likelihood* [23, p. 86] in pattern classification, and thus we are only concerned with the length between two bounds of the log-likelihood interval, i.e., $H(o) = |ln\overline{h}(o) - ln\underline{h}(o)|$. In Fig. 4.2a, the Gaussian primary MF with uncertain mean has

$$H(o) = \begin{cases} 2k_m|o - \mu|/\sigma, & o \leq \mu - k_m\sigma, \ o \geq \mu + k_m\sigma; \\ |o - \mu|^2/2\sigma^2 + k_m|o - \mu|/\sigma + k_m^2/2, & \mu - k_m\sigma < o < \mu + k_m\sigma. \end{cases} \tag{4.12}$$

In Fig. 4.2b, the Gaussian primary MF with uncertain std has

$$H(o) = (1/k_v^2 - k_v^2)|o - \mu|^2/2\sigma^2. \tag{4.13}$$

In (4.12) and (4.13), $\mu$ and $\sigma$ are the mean and std of the original certain T1 MF without uncertainty. Both (4.12) and (4.13) are increasing functions in terms of the deviation $|o - \mu|$. For example, given a fixed $k_m$, the farther the $o$ deviates from $\mu$, the larger $H(o)$ is in (4.12), which reflects a higher extent of the likelihood

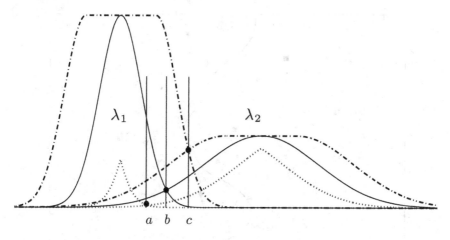

**Fig. 4.3** The decision rule based on *lower* and *upper* boundaries

uncertainty. This relationship accords with the outlier analysis. If the outlier $o$ deviates farther from the center of the class-conditional distribution, it has a larger $H(o)$ showing its greater uncertainty to the class model. Therefore, the interval log-likelihood, $[\ln \overline{h}, \ln \underline{h}]$, can provide additional information to measure the observation uncertainty to the class-conditional distribution. In Sect. 4.3 we shall use the interval log-likelihood as input feature vectors to the GLM for pattern classification.

Training a T2 FGMM is to estimate parameters $\mu$, $\Sigma$, and the factor $k_m$ or $k_v$. In this chapter, the factors $k_m$ or $k_v$ are the constants according to prior knowledge, and thus parameter estimation of T2 FGMMs includes two steps.

1. Estimate GMMs parameters by the EM algorithm;
2. Add the factor $k_m$ or $k_v$ to GMMs to produce T2 FGMM-UMs or T2 FGMM-UVs.

Without loss of generality, we take the T2 FGMM-UM as an example to explain how to use it for density modeling and classification. First, we need to make the best classification decision based on interval likelihoods. For simplicity, let us consider the univariate case for two-category classification in Fig. 4.3. The solid lines represent class-conditional densities $p(o|\lambda)$ of class models $\lambda_1$ and $\lambda_2$. They intersect at point $b$, which is the best decision boundary in case of the equal prior probability according to Bayesian decision theory [23], that is, for minimum classification error rate:

$$\text{Decide } \lambda_1 \text{ if } p(o|\lambda_1) > p(o|\lambda_2). \tag{4.14}$$

Because of uncertain parameters, the class-conditional densities have the FOUs constrained by the lower and upper boundaries, which intersect at points $a$ and $c$, respectively. Similar to Bayesian decision theory, to minimize the overall classification error rate, we have the following decision rule:

$$\text{Decide } \lambda_1 \text{ if } \underline{h}(o|\lambda_1) > \underline{h}(o|\lambda_2) \text{ and } \overline{h}(o|\lambda_1) > \overline{h}(o|\lambda_2). \tag{4.15}$$

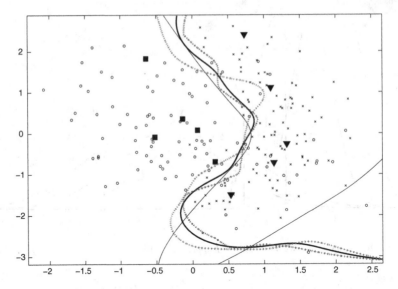

**Fig. 4.4** The *black squares* and *triangles* are *centers* of the mixture components. The *thin black solid line* is the optimal Bayesian decision boundary by the generating distributions. The *thick black solid line* is the decision boundary by GMMs. The *dotted lines* are the decision boundaries by the *lower* and *upper* likelihoods of T2 FGMM-UMs with $k_m = 1$

This rule classifies all samples left of the point $a$ to $\lambda_1$, and all samples right of the point $c$ to $\lambda_2$ in Fig. 3.10. Samples between $a$ and $c$ (outliers) are difficult to classify. If we classify them based on the lower boundary, the error rate decided by the upper boundary will increase. If we classify them based on the upper boundary, then in the meantime, we also increase the error rate by the lower boundary. Between $a$ and $c$, no decision boundary can reduce the error simultaneously. In practice, samples between $a$ and $c$ are often similar patterns that are difficult to classify. Most classification errors are made by these samples. Figure 4.4 shows the two decision boundaries by the lower and upper (dotted lines) boundaries of the T2 FGMM-UM with $k_m = 1$. These two boundaries twist around the decision boundary (thick black) of the GMM. The optimal decision boundary by Bayesian classifier [23] is the thin black line. It follows from Figs. 3.10 and 4.4 that a linear combination of lower and upper boundaries may be better than the decision boundary of the GMM in terms of the classification rate. Here, we propose the GLM [26] for the classification decision-making. The GLM automatically determines a weight matrix that linearly combines interval likelihoods based on training data. It weighs the upper and lower bounds of the interval likelihood and uses their sum as the output for decision-making. Indeed, the GLM assumes that the linear combination of the two bounds of the interval likelihood is better than the precise likelihood without uncertainty. Although this linear relationship may be not true for classification decision-making, it is simple and efficient in practice, and does not influence the uncertainty conveyed by the interval likelihoods.

**Fig. 4.5** The pattern classification system based on the T2 FGMM-UM. We adopt the GLM to make the classification decision from interval likelihoods

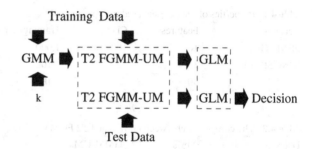

Given $1 \leq c \leq C$ categories, we first build a GMM for each category in turn, and estimate parameters on labeled data by the EM algorithm. Second, we add the factor $k_m$ to GMMs in (4.10) to produce T2 FGMM-UMs according to prior knowledge. Third, we evaluate each training sample by all T2 FGMM-UMs, and the output interval likelihoods compose a feature vector, $\mathbf{x} = [\underline{h}_1, \overline{h}_1, \ldots, \underline{h}_C, \overline{h}_C]$, for each training sample. To make classification decision from interval likelihoods, we adopt the GLM,

$$\mathbf{y} = f_a(\mathbf{w}\mathbf{x} + \mathbf{b}), \tag{4.16}$$

where $\mathbf{w}$ is the weight matrix, $\mathbf{b}$ is the bias vector, and $f_a$ is the logistic activation function. The iterated reweighted least square algorithm [26, Sect. 4.5] estimates the GLM parameters from all feature vectors $\mathbf{x}$ of training data. This estimated GLM is in turn used to classify the interval likelihood features with respect to test data. As a summary, Fig. 4.5 illustrates the pattern classification system based on the T2 FGMM-UM. Each feature vector of interval likelihoods has $2C$ length, where $C$ is the number of categories. In most pattern classification tasks, $C$ is not large so that $\mathbf{x}$ has a low dimensionality. When $C$ is large and training samples are insufficient, we have to reduce the $\mathbf{x}$'s dimensionality using principle component analysis (PCA) [23] in order to estimate the GLM parameters accurately.

## 4.3 Multi-category Pattern Classification

GMMs and EM algorithm have been efficiently implemented in the NETLAB toolbox of MATLAB© [26]. To test T2 FGMMs, we performed experiments on three datasets: PENDIGITS, IONOSPHERE, and WDBC from UCI [27] repository. All datasets have only numeric-valued features. In PENDIGITS dataset, we classified ten digits. For each digit, the original training data contains about 750 samples, from which we selected only 100 samples to train GMMs and T2 FGMMs. The total number of test samples is 3,498. In IONOSPHERE and WDBC datasets, we classified 251 and 469 test samples into two classes, respectively. From the remaining data, we randomly selected 50 samples for the training of each class. Because the amount

**Table 4.1**  Properties of the datasets used in experiments

| Datasets | Features | Classes | Training samples | Test samples |
|---|---|---|---|---|
| PENDIGITS | 16 | 10 | 1,000 | 3,498 |
| IONOSPHERE | 34 | 2 | 100 | 251 |
| WDBC | 30 | 2 | 100 | 469 |

**Table 4.2**  The comparison between GMMs and T2 FGMM-UMs

| Datasets | GMMs | T2 FGMM-UMs | | | |
|---|---|---|---|---|---|
| | | $k_m = 0.5$ | $k_m = 1.0$ | $k_m = 1.5$ | $k_m = 2.0$ |
| PENDIGITS | 88.3 | 89.7 | 90.2 | 91.4 | 91.9 |
| IONOSPHERE | 75.3 | 77.7 | 77.7 | 77.3 | 76.9 |
| WDBC | 93.6 | 94.9 | 94.0 | 93.8 | 93.6 |

**Table 4.3**  The comparison between GMMs and T2 FGMM-UVs

| Datasets | GMMs | T2 FGMM-UVs | | | |
|---|---|---|---|---|---|
| | | $k_v = 0.95$ | $k_v = 0.9$ | $k_v = 0.8$ | $k_v = 0.7$ |
| PENDIGITS | 88.3 | 88.7 | 89.2 | 90.6 | 90.5 |
| IONOSPHERE | 75.3 | 77.7 | 77.7 | 77.7 | 77.7 |
| WDBC | 93.6 | 93.8 | 94.2 | 94.0 | 94.5 |

of training samples is about four times smaller than that of test samples, training data are insufficient compared with the large amount of test data. Table 4.1 gives the properties of datasets we used in experiments. We used two mixture components in GMMs to model the class-conditional density. After parameter estimation, we added the factor $k_m$ (4.10) or $k_v$ (4.11) to produce T2 FGMM-UMs or T2 FGMM-UVs, and used the GLM to make the final classification decision in Fig. 4.5. At present, we have little prior knowledge to decide the best $k_m$ or $k_v$, so we tested a group of T2 FGMM-UMs with $k_m = 0.5, 1, 1.5, 2$, and a group of T2 FGMM-UVs with $k_v = 0.95, 0.9, 0.8, 0.7$. In this way, we constructed a group of T2 FGMM-UMs and T2 FGMM-UVs with different areas of the FOU, which may play different roles in pattern classification.

Tables 4.2 and 4.3 compare the performance of GMMs with those of T2 FGMM-UMs and T2 FGMM-UVs in the classification rate (%). Important findings are as follows.

1. T2 FGMM-UMs and T2 FGMM-UVs outperform GMMs in terms of the average classification rate, with 2.5 and 1.5 % higher for PENDIGITS, 2.1 and 2.4 % higher for IONOSPHER, and 0.5 and 0.5 % higher for WDBC for all vales of $k_m$ and $k_v$.
2. The classification rate is stable to the different values of $k_m$ and $k_v$. Its average std is 0.6 for T2 FGMM-UMs, and is 0.4 for T2 FGMM-UVs.

3. In PENDIGITS, the larger the $k_m$ or $k_v$, the higher the classification rate. In IONOSPHERE, the classification rate remains stable in different values of $k_m$ or $k_v$. In WDBC, the larger the $k_m$, the lower the classification rate, while the larger the $k_v$, the higher the classification rate.

From above findings, we observe that T2 FGMMs are consistently as good as or even better than GMMs in terms of the classification rate based on insufficient training data. The classification performance of the T2 FGMM is insensitive to different areas of the FOU, which provides convenience in designing practical systems. The larger the FOU in T2 FGMMs, the slightly better the performance. Intuitively, the better classification performance of T2 FGMMs is due to their ability to describe uncertainty of decision boundaries in Fig. 4.4. As we see, the decision boundaries (dotted lines) by lower and upper likelihoods of the T2 FGMM cover the optimal decision (thick solid line) by the GMM, and the linear combination of the two boundaries by the GLM may provide a better classification performance because the GLM encodes the information of interval likelihoods from training samples.

Although T2 FGMMs pay at least a doubled computational cost in the classification decision process to achieve a less than 3 % improvement in our experiments, they employ an enriched vocabulary to describe the uncertain parameters of the GMMs induced by insufficient or noisy data. For example, the FOU defines the bounds of the uncertain GMMs, and the secondary membership function directly measures the uncertain likelihoods of the uncertain GMMs. While other standard methods like C4.5 can achieve around 90 % precision for IONOSPHER dataset [28], they have not been tested under the condition of insufficient training data. Our results show that the FOU of the T2 FS at least enhances the robustness of the classical GMMs in case of insufficient training data. In the future work, we may improve T2 FGMMs by using different types of secondary membership functions because the area of the uniform FOU does not affect the classification performance very much.

# References

1. McLachlan, G.J., Basford, K.E.: Mixture Models: Inference and Applications to Clustering. Marcel Dekker, New York (1988)
2. Dempster, A.P., Laird, N.M., Rubin, D.B.: Maximum likelihood from incomplete data via the EM algorithm. J. R. Stat. Soc. Ser. B **39**, 1–38 (1977)
3. Mendel, J.M.: Uncertain Rule-based Fuzzy Logic Systems: Introduction and New Directions. Prentice-Hall, Upper-Saddle River (2001)
4. Mendel, J.M., John, R.I.B.: Type-2 fuzzy sets made simple. IEEE Trans. Fuzzy Syst. **10**(2), 117–127 (2002)
5. Mendel, J.M.: Type-2 fuzzy sets: some questions and answers. IEEE Connect. Newsl. IEEE Neural Netw. Soc. **1**, 10–13 (2003)
6. Mendel, J.M.: Advances in type-2 fuzzy sets and systems. Inf. Sci. **177**, 84–110 (2007)
7. Zeng, J., Liu, Z.Q.: Type-2 fuzzy hidden Markov models and their application to speech recognition. IEEE Trans. Fuzzy Syst. **14**(3), 454–467 (2006)
8. Zeng, J., Liu, Z.Q.: Type-2 fuzzy sets for pattern recognition: the state-of-the-art. J. Uncertain Syst. **1**(3), 163–177 (2007)

9. John, R.I.B.: Perception modeling using type-2 fuzzy sets. Ph.D. thesis, De Montfort University, Leicester (2000)
10. Liang, Q., Mendel, J.M.: MPEG VBR video traffic modeling and classification using fuzzy technique. IEEE Trans. Fuzzy Syst. **9**(1), 183–193 (2001)
11. Mitchell, H.: Pattern recognition using type-II fuzzy sets. Inf. Sci. **170**, 409–418 (2005)
12. Tizhoosh, H.R.: Image thresholding using type II fuzzy sets. Pattern Recognit. **38**(12), 2363–2372 (2005)
13. Wu, H., Mendel, J.M.: Classification of battlefield ground vehicles using acoustic features and fuzzy logic rule-based classifiers. IEEE Trans. Fuzzy Syst. **15**(1), 56–72 (2007)
14. Zeng, J., Xie, L., Liu, Z.Q.: Type-2 fuzzy Gaussian mixture models. Pattern Recognit. **41**(12), 3636–3643 (2008)
15. Karnik, N.N., Mendel, J.M.: Operations on type-2 fuzzy sets. Fuzzy Sets Syst. **122**, 327–348 (2001)
16. Liang, Q., Mendel, J.M.: Interval type-2 fuzzy logic systems: theory and design. IEEE Trans. Fuzzy Syst. **8**(5), 535–549 (2000)
17. Tran, D., VanLe, T., Wagner, M.: Fuzzy Gaussian mixture models for speaker recognition. In: Proceedings of the International Conference on Spoken Language Processing, Sydney, pp. 759–762 (1998)
18. Bezdek, J.C.: Pattern Recognition with Fuzzy Objective Function Algorithms. Plenum Press, New York (1981)
19. Caillol, H., Pieczynski, W., Hillion, A.: Estimation of fuzzy gaussian mixture and unsupervised statisticalimage segmentation. IEEE Trans. Image Process. **6**(3), 425–440 (1997)
20. Tran, D., Wagner, M.: Fuzzy approach to Gaussian mixture models and generalised Gaussian mixture models. In: Proceedings of the Computation Intelligence Methods and Applications, New York, pp. 154–158 (1999)
21. Bazi, Y., Bruzzone, L., Melgani, F.: Image thresholding based on the EM algorithm and the generalized Gaussian distribution. Pattern Recognit. **40**(2), 619–634 (2007)
22. Zarandia, M.H.F., Türksenb, I., Kasbi, O.T.: Type-2 fuzzy modeling for desulphurization of steel process. Expert Syst. Appl. **32**(1), 157–171 (2007)
23. Duda, R.O., Hart, P.E., Stork, D.G.: Pattern Classification, 2nd edn. Wiley, New York (2001)
24. Cover, T.M., Thomas, J.A.: Elements of Information Theory. Wiley, New York (1991)
25. Wu, H., Mendel, J.M.: Uncertainty bounds and their use in the design of interval type-2 fuzzy logic systems. IEEE Trans. Fuzzy Syst. **10**(5), 622–639 (2002)
26. Nabney, I.T.: NETLAB: Algorithms for Pattern Recognitions. Springer, London (2002)
27. Newman, D., Hettich, S., Blake, C., Merz, C.: UCI repository of machine learning databases (1998). http://www.ics.uci.edu/~mlearn/MLRepository.html
28. Freund, Y., Schapire, R.E.: Experiments with a new boosting algorithm. In: Proceedings of the International Conference on Machine Learning (ICML'96), Bari, pp. 148–156 (1996)

# Chapter 5
# Type-2 Fuzzy Hidden Moarkov Models

**Abstract** This chapter extends hidden Markov models (HMMs) to type-2 fuzzy HMMs (T2 FHMMs). We derive the T2 fuzzy forward-backward algorithm and Viterbi algorithm using T2 FS operations. To investigate the effectiveness of T2 FHMMs, we apply them to phoneme classification and recognition on the TIMIT speech database. Experimental results show that T2 FHMMs can effectively handle noise and dialect uncertainties in speech signals besides a better classification performance than the classical HMMs. We also find that the larger area of the FOU in T2 FHMMs with uncertain mean vectors performs better in classification when the signal-to-noise ratio is lower.

## 5.1 Hidden Markov Models

Now we focus on the problem of modeling sequential data at successive times, $1 \leq i \leq I$, by the first-order Markov model mentioned in Sect. 2.2. We continue to assume each label is a hidden variable that can generate observations randomly. So this first-order Markov model is called the hidden Markov model (HMM) for the hidden labels. Because of simple mathematical form and efficient inference algorithms, HMMs have found greatest use in speech recognition, gesture recognition, and time series prediction.

In an HMM, the transition probability in Eq. (2.3) is denoted by $a_{jj'}$,

$$a_{jj'} = P(s_i = j'|s_{i-1} = j), \tag{5.1}$$

which is the time-independent probability of having label $j'$ at time $i$ given that the label at time $i-1$ was $j$. The transition probability satisfy

$$\sum_{j'=1}^{J} a_{jj'} = 1, \ 1 \leq j, j' \leq J. \tag{5.2}$$

In the so-called left-right HMM, we often constrain that no transitions are allowed to labels whose indices are lower than the current label. For instance, label 1 can

© Tsinghua University Press, Beijing and Springer-Verlag Berlin Heidelberg 2015
J. Zeng and Z.-Q. Liu, *Type-2 Fuzzy Graphical Models for Pattern Recognition*,
Studies in Computational Intelligence 591, DOI 10.1007/978-3-662-44690-4_5

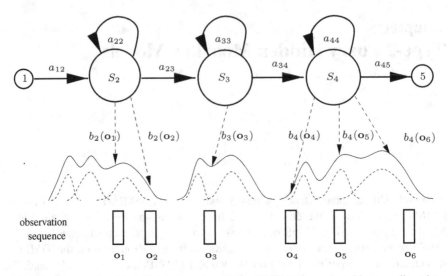

**Fig. 5.1** An example of the left-right Gaussian mixture HMM, where no transitions are allowed to states whose indices are lower than the current state

jump to 2 and 3, whereas 2 cannot jump back to 1 and 3 cannot jump back to 2 and 1. Without loss of generality, we assume that the generating distributions of hidden labels are GMMs denoted by $b_j(\mathbf{o}_i)$, and from Eq. (2.14) we obtain

$$b_j(\mathbf{o}_i) = p(\mathbf{o}_i|s_i = j) = \sum_{m=1}^{M_s} w_{jm} G(\mathbf{o}_i; \boldsymbol{\mu}_{jm}, \boldsymbol{\Sigma}_{jm}), \tag{5.3}$$

where $M_s$ is the number of mixture components, $w_{jm}$ is the weight of the $m$th component, and $G(; \boldsymbol{\mu}, \boldsymbol{\Sigma})$ is a multivariate Gaussian with mean vector $\boldsymbol{\mu}$ and covariance matrix $\boldsymbol{\Sigma}$,

$$G(\mathbf{o}; \boldsymbol{\mu}; \boldsymbol{\Sigma}) = \frac{1}{\sqrt{(2\pi)^d|\boldsymbol{\Sigma}|}} e^{-\frac{1}{2}(\mathbf{o}-\boldsymbol{\mu})'\boldsymbol{\Sigma}^{-1}(\mathbf{o}-\boldsymbol{\mu})}, \tag{5.4}$$

where $d$ is the dimensionality of $\mathbf{o}$. In speech recognition, to connect many HMMs for a word or sentence, the first and the final labels are non-generating in a $J$-label HMM. Thus, an HMM is defined by the parameter set, $\lambda = \{a_{jj'}, b_j(\mathbf{o}_i), j, j' = 1, \ldots, J\}$. Figure 5.1 shows the left-right HMMs.

With these preliminaries behind us, we can now focus on the three central issues in HMMs:

1. The Evaluation Problem: Suppose we have an HMM $\lambda$. Determine the probability $P(\mathbf{O}|\lambda)$ in Eq. (2.8), which means the probability that a particular sequence of observations $\mathbf{O}$ was generated by that model.

2. The Decoding Problem: Suppose we have an HMM $\lambda$ as well as a set observations **O**. Determine the most likely labeling configuration $\mathscr{S}^*$ in Eq. (2.10) that led to those observations.
3. The Learning Problem: Suppose we are given the topology of an HMM (the number of labels and the number of mixtures) but not the parameters $\lambda$. Given a set of training observations $\mathbf{O}^r$, $1 \leq r \leq R$, determine these parameters $\lambda$.

We consider each of these problems in turn.

### 5.1.1 The Forward-Backward Algorithm

Due to the causal neighborhood system of the HMM, the evaluation problem can be solved by a dynamic programming method called the forward-backward algorithm, whose computational complexity is $O(J^2 I)$.

We define the forward variable as

$$\alpha_j(i) = P(\mathbf{o}_1, \mathbf{o}_2, \ldots, \mathbf{o}_i, s_i = j | \lambda), \tag{5.5}$$

which is the joint probability of the partial observation sequence, $\{\mathbf{o}_1, \ldots, \mathbf{o}_i\}$, (until time $i$), and label $j$ at time $i$, given the HMM $\lambda$. Then we define the backward variable as

$$\beta_j(t) = P(\mathbf{o}_{i+1}, \mathbf{o}_{i+2}, \ldots, \mathbf{o}_I | s_i = j, \lambda), \tag{5.6}$$

which is the conditional probability of the partial observation sequence from time $i + 1$ to the end, given label $j$ at time $i$ and the HMM $\lambda$. After recursions for time $i$, we can get $\alpha_j(I)$ and $\beta_1(1)$, which are actually $P(\mathbf{O}|\lambda)$ from the definitions of forward and backward variables.

The forward probability is a joint probability, whereas the backward probability is a conditional probability. This allows the joint probability of the label $j$ and observations **O** to be determined by taking the product of forward and backward probabilities,

$$P(s_i = j, \mathbf{O}|\lambda) = \alpha_j(i)\beta_j(i). \tag{5.7}$$

If $L_j(i)$ denotes the posterior probability of the label $j$ at time $i$ given observations **O**, from Eq. (5.7) we obtain

$$\begin{aligned}
L_j(i) &= P(s_i = j | \mathbf{O}, \lambda) \\
&= \frac{P(\mathbf{O}, s_i = j | \lambda)}{P(\mathbf{O}|\lambda)} \\
&= \frac{\alpha_j(i)\beta_j(i)}{P(\mathbf{O}|\lambda)}.
\end{aligned} \tag{5.8}$$

```
input   : O = {o₁, . . . , o_I}, λ = {a_{jj'}, b_j(o_i)}.
output  : α_j(i), β_j(i), P(O|λ), L_j(i).
initialize: α₁(1) ← 1, α_{j'}(1) ← a_{1j'}b_{j'}(o₁), β_j(I) ← a_{jN}, 2 ≤ j, j' ≤ J − 1.
1  begin
2         for i ← 2 to I do
3               for j' ← 2 to J − 1, j ← 2 to J − 1 do
4                     α_{j'}(i) ← [∑_{j=2}^{J−1} α_j(i − 1)a_{jj'}] b_{j'}(o_i);
5                     β_j(I − i + 1) ← ∑_{j'=2}^{J−1} a_{jj'}b_{j'}(o_{I−i+2})β_{j'}(I − i + 2);
6               end
7         end
8         α_J(I) ← ∑_{j=2}^{J−1} α_j(I)a_{jJ};
9         β₁(1) ← ∑_{j'=2}^{J−1} a_{1j'}b_{j'}(o₁)β_{j'}(1);
10        P(O|λ) ← α_J(I) = β₁(1);
11        L_j(i) ← (α_j(i)β_j(i)) / P(O|λ);
12 end
```

**Fig. 5.2** The forward-backward algorithm

This posterior probability is the weight that each observation vector $o_i$ contributes to estimate the maximum-likelihood parameter values for each hidden label $j$. We will use this posterior probability to solve the learning problem by the Baum–Welch algorithm. The forward-backward algorithm is shown in Fig. 5.2. Note that the labels 1 and $J$ are non-generating.

### 5.1.2 The Viterbi Algorithm

The decoding problem is to find the best labeling configuration $\mathscr{S}^*$ and the joint probability $P(\mathscr{S}^*, O|\lambda)$ given an HMM $\lambda$. We define the variable as

$$\phi_j(i) = \max_{s_1,...,s_{i-1}} P(s_1, . . . , s_i = j, o_1, . . . , o_i|\lambda), \qquad (5.9)$$

which is the highest joint probability for the first $i$ observations and ends in label $j$ along a single path. The Viterbi algorithm can solve $\phi_j(i)$ recursively, which is almost the same with the forward-backward algorithm except that the summation operator is replaced by a maximum operator at each time step $i$. So the computational complexity of Viterbi algorithm is also $O(J^2 I)$. By recursions for time $i$, we can also retrieve the best labeling configuration as shown in Fig. 5.3, where $\psi_i(j)$ is used to keep track of the best label for each $i$ and $j$.

```
input   : O = {o₁, ..., o_I}, λ = {a_{jj'}, b_j(o_i)}.
output  : P(S*, O|λ), S* = {s*₁, ..., s*_I}.
initialize: φ₁(1) ← 1, φ_{j'}(1) ← a_{1j'} b_{j'}(o₁), ψ₁(j) ← 1, 2 ≤ j, j' ≤ J − 1.
1  begin
2      for i ← 2 to I do
3          for j' ← 2 to J − 1 do
4              φ_{j'}(i) ← [max_{j=2}^{J−1} φ_j(i − 1)a_{jj'}] b_{j'}(o_i);
5              ψ_i(j') ← arg max_{j=2}^{J−1} φ_j(i − 1)a_{jj'};
6          end
7      end
8      φ_J(I) ← max_{j=2}^{J−1} φ_j(I)a_{jJ};
9      P(S*, O|λ) ← φ_J(I);
10     s*_I ← arg max_{j=2}^{J−1} φ_j(I)a_{jJ};
11     for i ← I to 2 do
12         s*_{i−1} ← ψ_i(s*_i);
13     end
14 end
```

**Fig. 5.3** The Viterbi algorithm

## 5.1.3 The Baum–Welch Algorithm

The goal in the HMM learning is to determine model parameters—the transition probability $a_{jj'}$ and the generating distribution $b_j(o_i)$—from an ensemble of training samples $O^r$, $1 \leq r \leq R$. We ca use a generalized expectation-maximization (EM) algorithm, called the Baum–Welch algorithm, to iteratively update the model parameters according to the maximum-likelihood criterion. The Baum–Welch algorithm can repeat meany times until the average $P_r(O^r|\lambda)$ for all training samples is not higher than the value at the previous iteration. Alternatively, we can end the learning process by limiting the maximum iteration times $T_{max}$. On each iteration, the Baum–Welch algorithm demands merely that parameters should be improved not optimized. Figure 5.4 shows the Baum–Welch algorithm.

In practice, we use the k-means clustering and the Viterbi algorithm to initialize an HMM given its topology, and refine the HMM's parameters by the Baum–Welch algorithm. Through the Viterbi decoding, the best labeling configuration $\mathscr{S}^*$ implies an alignment of training observation vectors with labels. Within each label, a further alignment of observation vectors to mixture component of GMMs is made by the highest likelihood criterion. Finally, we initialize GMMs' parameters by the observation vectors within each mixture component. The initialization algorithm can repeat meany times until the average $P(O^r, \mathscr{S}^*|\lambda)$ for all training samples is not higher than the value at the previous iteration, or the maximum iteration times $T_{max}$ is achieved. The initialization algorithm is shown in Fig. 5.5.

To get accurate HMM, a large amount of training samples is needed. When the amount of training samples is small, certain mixture components will have very little associated training data, so either the variances or the corresponding mixture weight becomes very small. If either of these events happen, the mixture component is deleted, provided that at least one component in that label is left. Because of

> **input** : $\mathbf{O}^r = \{\mathbf{o}_1^r, \ldots, \mathbf{o}_I^r\}, 1 \leq r \leq R, \lambda = \{a_{jj'}, b_j(\mathbf{o}_i)\}$.
> **output** : $\hat{\lambda} = \{\hat{a}_{jj'}, \hat{b}_j(\mathbf{o}_i)\}$.
>
> **1** **begin**
> **2**      **for** $r \leftarrow 1$ **to** $R$ **do**
> **3**      |    $Pr(\mathbf{O}^r|\lambda), \alpha_j^r(i), \beta_j^r(i), L_j^r(i) \leftarrow \texttt{forwardbackward}(\mathbf{O}^r, \lambda)$;
> **4**      **end**
>
> **5**      $\hat{a}_{1j'} \leftarrow \dfrac{1}{R} \sum_{r=1}^{R} \dfrac{\alpha_{j'}^r(1)\beta_{j'}^r(1)}{Pr}$;
>
> **6**      $\hat{a}_{jj'} \leftarrow \dfrac{\sum_{r=1}^{R} \frac{1}{Pr} \sum_{i=1}^{I_r-1} \alpha_j^r(i) a_{jj'} b_{j'}(\mathbf{o}_{i+1}^r)\beta_{j'}^r(i+1)}{\sum_{r=1}^{R} \frac{1}{Pr} \sum_{i=1}^{I_r} \alpha_j^r(i)\beta_j^r(i)}$;
>
> **7**      $\hat{a}_{jJ} \leftarrow \dfrac{\sum_{r=1}^{R} \frac{1}{Pr} \alpha_j^r(I)\beta_j^r(I)}{\sum_{r=1}^{R} \frac{1}{Pr} \sum_{i=1}^{I_r} \alpha_j^r(i)\beta_j^r(i)}$;
>
> **8**      $\hat{\mu}_{jm} \leftarrow \dfrac{\sum_{r=1}^{R} \sum_{i=1}^{I_r} L_{jm}^r(i)\mathbf{o}_i^r}{\sum_{r=1}^{R} \sum_{i=1}^{I_r} L_{jm}^r(i)}$;
>
> **9**      $\hat{\Sigma}_{jm} \leftarrow \dfrac{\sum_{r=1}^{R} \sum_{i=1}^{I_r} L_{jm}^r(i)(\mathbf{o}_i^r - \hat{\mu}_{jm})(\mathbf{o}_i^r - \hat{\mu}_{jm})'}{\sum_{r=1}^{R} \sum_{i=1}^{I_r} L_{jm}^r(i)}$;
>
> **10**      $\hat{w}_{jm} \leftarrow \dfrac{\sum_{r=1}^{R} \sum_{i=1}^{I_r} L_{jm}^r(i)}{\sum_{r=1}^{R} \sum_{i=1}^{I_r} L_j^r(i)}$;
>
> **11** **end**

**Fig. 5.4** The Baum–Welch algorithm

limited training samples, covariance matrix $\Sigma$ may be singular and irreversible. In this case, the learning algorithm updates only mean vectors, and leaves the covariance matrix unchanged. Around 2–5 cycles of the Baum–Welch learning are normally sufficient, because repeated training to convergence may take a long time, and may lead to overfitting for the models that can become too closely matched to the training samples, and fail to generalize well to unknown test samples.

## 5.2 Type-2 Fuzzy Hidden Markov Models

HMMs have been used successfully in many applications, notably in speech recognition. Rabiner[1] shows three inherent limitations of HMMs for practical use: (1) the independence assumption, (2) the limited expressive power, and (3) the first-order Markov assumption. Given sufficient training data, the HMM can accurately represent the training data according to the ML criterion. However, the HMM may generalize poorly to unknown test data because of noise, insufficient training data, and incomplete information. Therefore, modeling these uncertainties is important in both the HMM and speech data. Figure 3.9 illustrates the importance of modeling uncertainty to achieve a better generalization to test data. Here, we enhance the HMM's expressive power for uncertainty using T2 FSs.

---

**input** : $\mathbf{O}^r = \{\mathbf{o}_1^r, \ldots, \mathbf{o}_I^r\}, 1 \leq r \leq R, J, M_s$.

**output** : $\hat{\lambda} = \{\hat{a}_{jj'}, \hat{b}_j(\mathbf{o}_i)\}$.

**initialize**: $\Theta_{jm}^r(i) \leftarrow 0$.

1 **begin**

2     $\boldsymbol{\mu}_{jm}, \boldsymbol{\Sigma}_{jm}, w_{jm} \leftarrow \texttt{kmeans}(\mathbf{O}^r, J, M_s), \forall r$;

3     **for** $r \leftarrow 1$ **to** $R$ **do**

4        $P(\mathbf{O}^r, \mathcal{S}^*|\lambda), \mathcal{S}^* \leftarrow \texttt{viterbi}(\mathbf{O}^r, \lambda)$;

5        **if** $s_i^r = label\ j\ mixture\ m$ **then**

6           $\Theta_{jm}^r(i) \leftarrow 1$;

7        **end**

8        $A_{jj'} \leftarrow$ total number of transitions from label $j$ to label $j'$;

9     **end**

10     $\hat{a}_{jj'} \leftarrow \dfrac{A_{jj'}}{\sum_{k=1}^{J} A_{ik}}$;

11     $\hat{\boldsymbol{\mu}}_{jm} \leftarrow \dfrac{\sum_{r=1}^{R} \sum_{i=1}^{I_r} \Theta_{jm}^r(i) \mathbf{o}_i^r}{\sum_{r=1}^{R} \sum_{i=1}^{I_r} \Theta_{jm}^r(i)}$;

12     $\hat{\boldsymbol{\Sigma}}_{jm} \leftarrow \dfrac{\sum_{r=1}^{R} \sum_{i=1}^{I_r} \Theta_{jm}^r(i)(\mathbf{o}_i^r - \hat{\boldsymbol{\mu}}_{jm})(\mathbf{o}_i^r - \hat{\boldsymbol{\mu}}_{jm})'}{\sum_{r=1}^{R} \sum_{i=1}^{I_r} \Theta_{jm}^r(i)}$;

13     $\hat{w}_{jm} \leftarrow \dfrac{\sum_{r=1}^{R} \sum_{i=1}^{I_r} \Theta_{jm}^r(i)}{\sum_{r=1}^{R} \sum_{i=1}^{I_r} \sum_{m=1}^{M_s} \Theta_{jm}^r(i)}$;

14 **end**

---

**Fig. 5.5** The initialization algorithm

In light of the T2 FS framework, let us reexamine the HMM's uncertainty: The parameters of the HMM are uncertain because of insufficient and corrupted training data. Meanwhile, such variations of speech data as noise and dialect greatly degrade the HMM's expressive power to evaluate the unknown test data. In view of these problems, we extend the classical HMM to the T2 FHMMs using T2 FSs. Because the T2 MF is three-dimensional, we may use one dimension, the primary MF, to evaluate the *likelihood* of the random data and use the other dimension, the secondary MF, to describe the *fuzziness* of that likelihood. Therefore, T2 FS is natural to handle two kinds of uncertainties, namely, randomness and fuzziness; Probability theory is associated with the former and FS theory is associated with the latter. From this perspective, the T2 FHMM is a hybrid scheme that allows some degree of fuzziness in the HMM's description. Just as the T2 FS is represented by a set of embedded T1 FSs, the T2 FHMM can be thought of being embedded with a set of HMMs. On the other hand, the T2 nonsingleton fuzzification is able to represent the uncertainty of data as T2 fuzzy vectors. Hence, both uncertainties of the HMM and speech data can be accounted for in the T2 FHMM framework by mapping uncertain input data to a T2 FS through set operations. Compared with the HMM, the T2 FHMM handles a sequence of T2 fuzzy vectors rather than a crisp numeric sequence, and the output of the T2 FHMM is an uncertain T1 FS rather than a crisp scalar.

**Fig. 5.6** Transition
probability is modeled by the
triangular T1 number

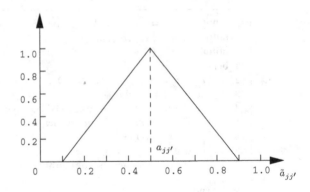

### 5.2.1 Elements of a Type-2 FHMM

Instead of observation probability density $b_j(\mathbf{o}_i)$, each fuzzy hidden label is associated with a T2 MF $\tilde{b}_j(\mathbf{o}_i)$ that is the membership of $\mathbf{o}_i$ to the fuzzy label $\tilde{s}_i = j$. We use the T2 FGMMs (4.3) or (4.4) to represent $\tilde{b}_j(\mathbf{o}_i)$. To model the uncertainty of the transition probability from labels $j$ to $j'$, we use the T1 fuzzy number $\tilde{a}_{jj'}$ with support $a_{jj'}$ (e.g., triangular as shown in Fig. 5.6). We assume that the first and final labels are non-generating in a $J$-label T2 FHMM.

The membership grade of $\tilde{a}_{jj'}$ reflects the uncertainty of the transition probability from $j$ to $j'$. The primary membership of $\tilde{b}_j(\mathbf{o}_i)$ is the membership that $\mathbf{o}_i$ belongs to the fuzzy label $\tilde{s}_i = j$, and the secondary grade reflects our belief to this membership. Let $\tilde{\Omega}_1 \times \tilde{\Omega}_2 \times \cdots \times \tilde{\Omega}_I$ denote the space of fuzzy observation sequence from time 1 to time $I$. The observation vector $\mathbf{o}_i$ is fuzzified as a vector of T2 fuzzy numbers denoted by the T2 MF $h_{\tilde{\Omega}}(\mathbf{o}_i)$. Given an observation sequence, $\mathbf{O} = \{\mathbf{o}_1, \mathbf{o}_2, \ldots, \mathbf{o}_I\}$, and a T2 FHMM $\tilde{\lambda}$, $h_{\tilde{\lambda}}(\mathbf{O})$ denotes the membership grade that the sequence $\mathbf{O}$ belongs to the T2 FHMM, and is a T1 FS that contains uncertain information conveyed by both $\tilde{b}_j(\mathbf{o}_i)$ and $\tilde{a}_{jj'}$.

With these preliminaries behind us, we now focus on deriving T2 fuzzy forward-backward algorithm, T2 fuzzy Viterbi algorithm, and Baum–Welch algorithm in turn.

### 5.2.2 The Type-2 Fuzzy Forward-Backward Algorithm

We define the T2 fuzzy forward variable as

$$\tilde{\alpha}_j(i) = h_{\tilde{\lambda}}(\mathbf{o}_1, \mathbf{o}_2, \ldots, \mathbf{o}_i, \tilde{s}_i = j), \tag{5.10}$$

i.e., the membership grade of the partial observation sequence, $\mathbf{o}_1, \ldots, \mathbf{o}_i$, (until time $i$) and label $j$ at time $i$, given the T2 FHMM $\tilde{\lambda}$. The T2 fuzzy backward variable is defined as

$$
\begin{array}{ll}
\textbf{input} & : \mathbf{O} = \{o_1, \ldots, o_I\}, \lambda = \{\tilde{a}_{jj'}, \tilde{b}_j(o_i), h_{\tilde{\Omega}}(o_i)\}. \\
\textbf{output} & : h_{\tilde{\lambda}}(\mathbf{O}), \tilde{\alpha}_j(i), \tilde{\beta}_j(i), L_j(i). \\
\textbf{initialize:} & \tilde{\alpha}_1(1) \leftarrow 1, \tilde{\alpha}_{j'}(1) \leftarrow \tilde{a}_{1j'} \sqcap [\sqcup_{o_1 \in \tilde{\Omega}}(h_{\tilde{\Omega}}(o_1) \sqcap \tilde{b}_{j'}(o_1))], \tilde{\beta}_j(I) \leftarrow \\
& \tilde{a}_{jJ}, \ 2 \le j, j' \le J - 1.
\end{array}
$$

```
 1  begin
 2      for i ← 2 to I do
 3          for j' ← 2 to J − 1, j ← 2 to J − 1 do
 4              α̃_{j'}(i) ←
                [⊔_{j=2}^{J−1}(α̃_j(i − 1) ⊓ ã_{jj'})] ⊓ [⊔_{o_i∈Ω̃}(h_{Ω̃}(o_i) ⊓ b̃_j(o_i))];
 5              β̃_j(I − i + 1) ← ⊔_{j'=2}^{J−1}{ã_{jj'} ⊓ [⊔_{o_{i−I+2}∈Ω̃}(h_{Ω̃}(o_{i−I+2}) ⊓
                b̃_{j'}(o_{I−i+2}))] ⊓ β̃_{j'}(I − i + 2)};
 6          end
 7      end
 8      α_J(I) ← ⊔_{j=2}^{J−1}(α̃_j(I) ⊓ ã_{jJ});
 9      β_1(1) ← ⊔_{j'=2}^{J−1}{ã_{1j'} ⊓ [⊔_{o_1∈Ω̃}(h_{Ω̃_1}(o_1) ⊓ b̃_{j'}(o_1))] ⊓ β̃_{j'}(1)};
10      h_{λ̃}(O) ← α̃_J(I) = β̃_1(1);
11      L_j(i) ← (1/𝔇(h_{λ̃}(O))) 𝔇(α̃_j(i) ⊓ β̃_j(i));
12  end
```

**Fig. 5.7** The type-2 fuzzy forward-backward algorithm

$$
\tilde{\beta}_j(i) = h_{\tilde{\lambda}}(o_{i+1}, o_{i+2} \ldots, o_I | \tilde{s}_i = j), \tag{5.11}
$$

i.e., the membership grade of the partial observation sequence from $i + 1$ to the end, given label $j$ at time $i$ and the T2 FHMM $\tilde{\lambda}$. From the definition of $\tilde{\alpha}_j(i)$, we have the total membership grade of $\mathbf{O}$ to the T2 FHMM $\tilde{\lambda}$ as follows,

$$
h_{\tilde{\lambda}}(\mathbf{O}) = \tilde{\alpha}_J(I). \tag{5.12}
$$

Note that the T2 fuzzy forward and backward variables allow the total membership grade of observation sequence being in the $j$th label at time $i$ to be determined by taking the meet operation "$\sqcap$" of them,

$$
\tilde{\alpha}_j(i) \sqcap \tilde{\beta}_j(i) = h_{\tilde{\lambda}}(\mathbf{O}, \tilde{s}_i = j), \tag{5.13}
$$

because $\tilde{\alpha}_j(i)$ accounts for the partial observation sequence $o_1, \ldots, o_i$, and $\tilde{\beta}_j(i)$ accounts for the remainder of the observation $o_{i+1}, o_{i+2}, \ldots, o_I$ given state label $\tilde{s}_i = j$. Let $\mathfrak{D}(A)$ denote the defuzzified value of a T1 FS $A$, which is a mapping from a T1 FS to a crisp scalar. The defuzzified membership grade of observation $o_i$ being in state $\tilde{s}_i = j$, denoted by $L_j(i)$, is

$$
L_j(i) = \frac{1}{\mathfrak{D}(h_{\tilde{\lambda}}(\mathbf{O}))} \mathfrak{D}(\tilde{\alpha}_j(i) \sqcap \tilde{\beta}_j(i)), \tag{5.14}
$$

where $\mathfrak{D}(h_{\tilde{\lambda}}(\mathbf{O}))$ is a normalization factor. The T2 fuzzy forward-backward algorithm is shown in Fig. 5.7.

$$
\begin{aligned}
&\textbf{input} \quad : \mathbf{O} = \{\mathbf{o}_1, \ldots, \mathbf{o}_I\}, \tilde{\lambda} = \\
&\qquad \{[\underline{a}_{jj'}, \overline{a}_{jj'}], [\underline{b}_j(i), b_j(i)], [\underline{h}_{\tilde{\Omega}}(\mathbf{o}_i), \overline{h}_{\tilde{\Omega}}(\mathbf{o}_i)]\}. \\
&\textbf{output} : [\underline{h}_{\tilde{\lambda}}(\mathbf{O}), h_{\tilde{\lambda}}(\mathbf{O})]. [\underline{\alpha}_j(i), \overline{\alpha}_j(i)]. [\underline{\beta}_j(i), \overline{\beta}_j(i)], L_j(i). \\
&\textbf{initialize:} \underline{\alpha}_1(1), \alpha_1(1) \leftarrow 1, \\
&\qquad \underline{\alpha}_{j'}(1) \leftarrow \underline{a}_{1j'} \star [\sup_{\mathbf{o}_1 \in \tilde{\Omega}} (\underline{h}_{\tilde{\Omega}}(\mathbf{o}_1) \star \underline{b}_{j'}(\mathbf{o}_1))], \overline{\alpha}_{j'}(1) \leftarrow \\
&\qquad \overline{a}_{1j'} \star [\sup_{\mathbf{o}_1 \in \tilde{\Omega}} (\overline{h}_{\tilde{\Omega}}(\mathbf{o}_1) \star b_{j'}(\mathbf{o}_1))], \underline{\beta}_j(I) \leftarrow \underline{a}_{jJ}, \overline{\beta}_j(I) \leftarrow \\
&\qquad \overline{a}_{jJ}, 2 \le j, j' \le J - 1.
\end{aligned}
$$

1  **begin**
2      **for** $i \leftarrow 2$ **to** $I$ **do**
3          **for** $j' \leftarrow 2$ **to** $J - 1, j \leftarrow 2$ **to** $J - 1$ **do**
4              $\underline{\alpha}_{j'}(i) \leftarrow$
            $[\sum_{j=2}^{J-1} (\underline{\alpha}_j(i-1) \star \underline{a}_{jj'})] \star [\sup_{\mathbf{o}_i \in \tilde{\Omega}} (\underline{h}_{\tilde{\Omega}}(\mathbf{o}_i) \star \underline{b}_j(\mathbf{o}_i))];$
5              $\overline{\alpha}_{j'}(i) \leftarrow$
            $[\sum_{j=2}^{J-1} (\overline{\alpha}_j(i-1) \star \overline{a}_{jj'})] \star [\sup_{\mathbf{o}_i \in \tilde{\Omega}} (\overline{h}_{\tilde{\Omega}}(\mathbf{o}_i) \star \overline{b}_j(\mathbf{o}_i))];$
6              $\underline{\beta}_j(I - i + 1) \leftarrow \sum_{j=2}^{J-1} \{\underline{a}_{jj'} \star [\sup_{\mathbf{o}_{I-i+2} \in \tilde{\Omega}} (\underline{h}_{\tilde{\Omega}}(\mathbf{o}_{I-i+2}) \star$
            $\underline{b}_{j'}(\mathbf{o}_{I-i+2})] \star \underline{\beta}_{j'}(I - i + 2)\};$
7              $\overline{\beta}_j(I - i + 1) \leftarrow \sum_{j=2}^{J-1} \{\overline{a}_{jj'} \star [\sup_{\mathbf{o}_{I-i+2} \in \tilde{\Omega}} (\overline{h}_{\tilde{\Omega}}(\mathbf{o}_{I-i+2}) \star$
            $\overline{b}_{j'}(\mathbf{o}_{I-i+2})] \star \overline{\beta}_{j'}(I - i + 2)\};$
8          **end**
9      **end**
10     $\underline{\alpha}_J(I) \leftarrow \sum_{j=2}^{J-1} (\underline{\alpha}_j(I) \star \underline{a}_{jJ});$
11     $\overline{\alpha}_J(I) \leftarrow \sum_{j=2}^{J-1} (\overline{\alpha}_j(I) \star \overline{a}_{jJ});$
12     $\underline{\beta}_1(1) \leftarrow \sum_{j'=2}^{N-1} \{\underline{a}_{1j'} \star [\sup_{\mathbf{o}_1 \in \tilde{\Omega}_1} (\underline{h}_{\tilde{\Omega}_1}(\mathbf{o}_1) \star \underline{b}_{j'}(\mathbf{o}_1))] \star \underline{\beta}_{j'}(1)\};$
13     $\overline{\beta}_1(1) \leftarrow \sum_{j'=2}^{N-1} \{\overline{a}_{1j'} \star [\sup_{\mathbf{o}_1 \in \tilde{\Omega}_1} (\overline{h}_{\tilde{\Omega}_1}(\mathbf{o}_1) \star \overline{b}_{j'}(\mathbf{o}_1))] \star \overline{\beta}_{j'}(1)\};$
14     $\underline{h}_{\tilde{\lambda}}(\mathbf{O}) \leftarrow \underline{\alpha}_J(I) = \underline{\beta}_1(1);$
15     $\overline{h}_{\tilde{\lambda}}(\mathbf{O}) \leftarrow \overline{\alpha}_J(I) = \overline{\beta}_1(1);$
16     $L_j(i) \leftarrow \frac{1}{\mathfrak{D}(h_{\tilde{\lambda}}(\mathbf{O}))} \mathfrak{D}(\tilde{\alpha}_j(i) \star \tilde{\beta}_j(i));$
17 **end**

**Fig. 5.8** The interval type-2 fuzzy forward-backward algorithm

The computation of the general T2 fuzzy forward-backward algorithm is prohibitive because the general T2 FS operations are complex. The IT2 FS is a special case of the general T2 FS, where all secondary grades equal to one so that the set operations are simplified to interval arithmetic. In the interval type-2 fuzzy HMMs (IT2 FHMMs), the fuzzy forward and backward variables are both interval sets, i.e., $\tilde{\alpha}_j(i) = [\underline{\alpha}_j(i), \overline{\alpha}_j(i)]$, $\tilde{\beta}_j(i) = [\underline{\beta}_j(i), \overline{\beta}_j(i)]$, and $\tilde{a}_{jj'} = [\underline{a}_{jj'}, \overline{a}_{jj'}]$. Thus $h_{\tilde{\lambda}}(\mathbf{O}) = [\underline{h}_{\tilde{\lambda}}(\mathbf{O}), \overline{h}_{\tilde{\lambda}}(\mathbf{O})]$ can be solved recursively by the IT2 fuzzy forward-backward algorithm as shown in Fig. 5.8, where $\star$ is the product $t$-norm, and all calculations are interval arithmetic.

### 5.2.3 The Type-2 Fuzzy Viterbi Algorithm

To find the best fuzzy labeling configuration $\tilde{\mathscr{S}}^*$ and the membership grade $h_{\tilde{\lambda}}(\tilde{\mathscr{S}}^*, \mathbf{O})$, we define the variable as

```
input   : O = {o₁, . . . , o_I}, λ = {ã_{jj'}, b̃_j(o_i), h_Ω̃(o_i)}.
output  : h_λ̃(S̃*, O), S̃* = {s̃₁*, . . . , s̃_I*}.
initialize: φ̃₁(1) ← 1, φ̃_{j'}(1) ← ã_{1j'} ⊓ [⊔_{o₁∈Ω̃}(h_Ω̃(o₁) ⊓ b̃_{j'}(o₁))], ψ̃₁(j) ←
            1, 2 ≤ j, j' ≤ J − 1.
 1  begin
 2  │   for i ← 2 to I do
 3  │   │   for j' ← 2 to J − 1 do
 4  │   │   │   φ̃_{j'}(i) ←
    │   │   │       ⊔_{j=2}^{J−1}{φ̃_j(i − 1) ⊓ ã_{jj'}} ⊓ [⊔_{o_i∈Ω̃}(h_Ω̃(o_i) ⊓ b̃_j(o_i))];
 5  │   │   │   ψ̃_i(j') ← arg max_{j=2}^{J−1} 𝔇(φ̃_j(i − 1) ⊓ ã_{jj'});
 6  │   │   end
 7  │   end
 8  │   φ̃_J(I) ← ⊔_{j=2}^{J−1}{φ̃_j(I) ⊓ ã_{jJ}};
 9  │   h_λ̃(S̃*, O) ← φ̃_J(I);
10  │   s̃_I* ← arg max_{j=2}^{J−1} 𝔇(φ̃_j(I) ⊓ ã_{jJ});
11  │   for i ← I to 2 do
12  │   │   s̃_{i−1}* ← ψ̃_i(s̃_i*);
13  │   end
14  end
```

**Fig. 5.9**  The type-2 fuzzy Viterbi algorithm

$$\tilde{\phi}_j(i) = \max_{\tilde{s}_1,\ldots,\tilde{s}_{i-1}} h_{\tilde{\lambda}}(\tilde{s}_1,\ldots,\tilde{s}_i = j, o_1,\ldots,o_i), \tag{5.15}$$

i.e., $\tilde{\phi}_j(i)$ is the best membership grade of the first $i$ observations and ends in label $j$ along a single fuzzy labeling configuration. By induction, we have the T2 fuzzy Viterbi algorithm as shown in Fig. 5.9, where $\tilde{\psi}_i(j)$ is used to keep track of the best fuzzy label for each $i$ and $j$. The IT2 fuzzy Viterbi algorithm is shown in Fig. 5.10, where $\tilde{\phi}_j(i) = [\underline{\phi}_j(i), \overline{\phi}_j(i)]$.

## 5.2.4 The Learning Algorithm

Suppose, a set of training observations $O^r$, $1 \leq r \leq R$, is used to estimate the parameters of a T2 FHMM with $M_s$ mixture components. In practice the T2 fuzzy Viterbi algorithm is used to initialize a T2 FHMM as shown in Fig. 5.11. Then we use the Baum–Welch algorithm to refine the parameters of the T2 FHMM as shown in Fig. 5.12.

The algorithms of parameter estimation for the IT2 FHMM are almost identical with Fig. 5.12, except that the defuzzified values are the centers of the intervals, and the $t$-norm "⋆" replaces the meet operator "⊓". Compared with the classical Baum–Welch algorithm in Fig. 5.4, the updating weight $L^r_{jm}(t)$ in Fig. 5.12 have many choices because of different defuzzification methods in Sect. 3.3.3. As far as the center of the interval is concerned, $L^r_{jm}(t)$ may be greater than that in the classical Baum–Welch when the training observation vector $o_i$ deviates far greater from the center of the underlying density, as shown in Fig. 5.13. This phenomenon reflects a big

$$
\begin{aligned}
&\textbf{input} \quad : \mathbf{O} = \{\mathbf{o}_1, \dots, \mathbf{o}_I\}, \tilde{\lambda} = \{\tilde{a}_{jj'}, \tilde{b}_j(\mathbf{o}_i)\}. \\
&\textbf{output} \quad : h_{\tilde{\lambda}}(\tilde{S}^*, \mathbf{O}) = [\underline{h}_{\tilde{\lambda}}(\underline{S}^*, \mathbf{O}), \overline{h}_{\tilde{\lambda}}(\overline{S}^*, \mathbf{O})], \tilde{S}^* = \{\underline{S}^*, \overline{S}^*\}. \\
&\textbf{initialize:} \; \phi_1(1) \leftarrow 1, \underline{\phi}_{j'}(1) \leftarrow \underline{a}_{1j'} \star [\sup_{\mathbf{o}_1 \in \tilde{\Omega}}(\underline{h}_{\tilde{\Omega}}(\mathbf{o}_1) \star \underline{b}_j(\mathbf{o}_1))], \underline{\psi}_1(j) \leftarrow \\
&\quad 1, \phi_1(1) \leftarrow 1, \overline{\phi}_{j'}(1) \leftarrow \\
&\quad a_{1j'} \star [\sup_{\mathbf{o}_1 \in \tilde{\Omega}}(\overline{h}_{\tilde{\Omega}}(\mathbf{o}_1) \star b_j(\mathbf{o}_1))], \overline{\psi}_1(j) \leftarrow 1, \; 2 \le j, j' \le J-1.
\end{aligned}
$$

```
1  begin
2     for i ← 2 to I do
3        for j' ← 2 to J − 1 do
4           φ_{j'}(i) ←
              max_{j=2}^{J-1} {φ_j(i − 1) ⋆ a_{jj'}} ⋆ [sup_{o_i∈Ω̃}(h_{Ω̃}(o_i) ⋆ b_j(o_i))];
5           ψ_i(j') ← arg max_{j=2}^{J-1} φ_j(i − 1) ⋆ a_{jj'};
6           φ̄_{j'}(i) ←
              max_{j=2}^{J-1} {φ̄_j(i − 1) ⋆ ā_{jj'}} ⋆ [sup_{o_i∈Ω̃}(h̄_{Ω̃}(o_i) ⋆ b̄_j(o_i))];
7           ψ̄_i(j') ← arg max_{j=2}^{J-1} φ̄_j(i − 1) ⋆ ā_{jj'};
8        end
9     end
10    φ_J(I) ← max_{j=2}^{J-1} φ_j(I) ⋆ a_{jJ};   s_I^* ← arg max_{j=2}^{J-1} φ_j(I) ⋆ a_{jJ};
11    φ̄_J(I) ← max_{j=2}^{J-1} φ̄_j(I) ⋆ ā_{jJ};   s̄_I^* ← arg max_{j=2}^{J-1} φ̄_j(I) ⋆ ā_{jJ};
12    for i ← I to 2 do
13       s_{i-1}^* ← ψ_i(s_i^*);   s̄_{i-1}^* ← ψ̄_i(s̄_i^*);
14    end
15    h_{λ̃}(S^*, O) ← φ_J(I), h̄_{λ̃}(S̄^*, O) ← φ̄_J(I);
16 end
```

**Fig. 5.10**  Interval type-2 fuzzy Viterbi algorithm

difference from the classical Baum-Welch: Though $\mathbf{o}_i$ may deviate from the center of the underlying density, it still affects parameter reestimation through a reasonable weight. In this sense the IT2 fuzzy forward-backward algorithm is a *softer* kind of generalized EM algorithm according to the ML criterion.

To apply the IT2 FHMM to real problems, we first determine the number of states and mixtures, and then we choose the form of the IT2 MF and fix the uncertainty factors $k_m$ or $k_v$ in Eqs. (4.3) and (4.4). At each recursion of the IT2 fuzzy forward-back algorithm and the Viterbi algorithm, the IT2 MF generates a continuous IT1 set that reflects the uncertainty of the primary grade. As shown in Fig. 5.11, we use the IT2 fuzzy Viterbi algorithm to initialize a prototype IT2 FHMM as follows.

1. Start by uniformly segmenting the data, associating each successive segment with successive labels, and further clustering data into mixtures by fuzzy c-means algorithm.
2. Initialize parameters by data with each mixture.
3. Produce a prototype IT2 FHMM by fixing uncertainty factors.
4. Use the IT2 fuzzy Viterbi algorithm to search the best state sequence $\mathscr{S}^*$ for all IT2 nonsingleton fuzzified observation sequences $\mathbf{O}^r, 1 \le r \le R$, which implies alignments of training sequences with the fuzzy labels.
5. Further associate the observation vector with the $m$th mixture having the largest defuzzified membership grade, i.e., $\arg\max_m\{[\sup_{\mathbf{o}_i \in \tilde{\Omega}}(\underline{h}_{\tilde{\Omega}}(\mathbf{o}_i) \star \underline{b}_{jm}(\mathbf{o}_i)) + \sup_{\mathbf{o}_i \in \tilde{\Omega}}(\overline{h}_{\tilde{\Omega}}(\mathbf{o}_i) \star \overline{b}_{jm}(\mathbf{o}_i))]/2\}$.

$$\begin{array}{ll}
& \textbf{input} \quad : \mathbf{O}^r = \{\mathbf{o}_1^r, \ldots, \mathbf{o}_I^r\}, 1 \leq r \leq R, J, M_s. \\
& \textbf{output} \quad : \hat{\lambda} = \{\hat{a}_{jj'}, \hat{b}_j(\mathbf{o}_i)\}. \\
& \textbf{initialize: } \Theta_{jm}^r(i) \leftarrow 0.
\end{array}$$

1 **begin**
2     $\boldsymbol{\mu}_{jm}, \boldsymbol{\Sigma}_{jm}, w_{jm} \leftarrow \texttt{fuzzy-c-means}(\mathbf{O}^r, J, M_s), \forall r;$
3     **for** $r \leftarrow 1$ **to** $R$ **do**
4        $h_{\tilde{\lambda}}(\tilde{S}^*, \mathbf{O}^r), \tilde{S}^* = \{\tilde{s}_1^*, \ldots, \tilde{s}_I^*\} \leftarrow \texttt{type2fuzzyviterbi}(\mathbf{O}^r, \lambda);$
5        **if** $s_i^{r*} = label\ j\ mixture\ m$ **then**
6          $\Theta_{jm}^r(i) \leftarrow 1;$
7        **end**
8        $A_{jj'} \leftarrow$ total number of transitions from label $j$ to label $j'$;
9     **end**

$$10 \quad \hat{a}_{jj'} \leftarrow \frac{A_{jj'}}{\sum_{k=1}^J A_{ik}};$$

$$11 \quad \hat{\mu}_{jm} \leftarrow \frac{\sum_{r=1}^R \sum_{i=1}^{I_r} \Theta_{jm}^r(i)\mathbf{o}_i^r}{\sum_{r=1}^R \sum_{i=1}^{I_r} \Theta_{jm}^r(i)};$$

$$12 \quad \hat{\Sigma}_{jm} \leftarrow \frac{\sum_{r=1}^R \sum_{i=1}^{I_r} \Theta_{jm}^r(i)(\mathbf{o}_i^r - \hat{\mu}_{jm})(\mathbf{o}_i^r - \hat{\mu}_{jm})'}{\sum_{r=1}^R \sum_{i=1}^{I_r} \Theta_{jm}^r(i)};$$

$$13 \quad \hat{w}_{jm} \leftarrow \frac{\sum_{r=1}^R \sum_{i=1}^{I_r} \Theta_{jm}^r(i)}{\sum_{r=1}^R \sum_{i=1}^{I_r} \sum_{m=1}^{M_s} \Theta_{jm}^r(i)};$$

14 **end**

**Fig. 5.11** The type-2 fuzzy initialization algorithm

6. Update the parameters by data (not fuzzified) with each mixture.
7. If the average membership grade of all training observations
$\frac{\sum_{r=1}^R [(\underline{h}_{\tilde{\lambda}}(\mathscr{S}^*, \mathbf{O}^r) + \overline{h}_{\tilde{\lambda}}(\mathscr{S}^*, \mathbf{O}^r))/2]}{R}$ for this iteration is not higher than the value at the previous iteration then stop, otherwise repeat steps 4–7 using the updated IT2 FHMM.

Finally, we use the Baum–Welch algorithm to refine the parameters of the initialized IT2 FHMM. As shown in Fig. 5.12, the steps to perform parameters reestimation are summarized as follows.

1. For every parameter vector/matrix requiring reestimation, allocate storage for the numerator and denominator summations. These storage are referred to as accumulators.
2. Calculate the IT2 fuzzy forward and backward variables of the IT2 nonsingleton fuzzified observation sequence $\mathbf{O}^r$, $1 \leq r \leq R$, for all labels $\tilde{s}_i = j$, mixtures $m$, and times $i$.
3. For each label $\tilde{s}_i = j$, mixture $m$, and time $i$, use the weight $L_{jm}^r(i)$ and the current observation $\mathbf{o}_i$ to update the accumulators for that mixture.

$$
\begin{array}{ll}
\textbf{input} & : \mathbf{O}^r = \{o_1^r, \ldots, o_I^r\}, 1 \le r \le R, \tilde{\lambda} = \{\tilde{a}_{jj'}, \tilde{b}_j(o_i)\}. \\[4pt]
\textbf{output} & : \hat{\lambda} = \{\hat{a}_{jj'}, \hat{b}_j(o_i)\}.
\end{array}
$$

**1** begin

**2**     for $r \leftarrow 1$ to $R$ do

**3**        $h_{\tilde{\lambda}}(\mathbf{O}^r), \tilde{\alpha}_j^r(i), \tilde{\beta}_j^r(i), L_j^r(i) \leftarrow \texttt{type2fuzzyforwardbackward}(\mathbf{O}^r, \tilde{\lambda})$;

**4**     end

**5**     $\hat{a}_{1j'} = \dfrac{1}{R} \sum_{r=1}^{R} \dfrac{1}{\mathfrak{D}(h_{\tilde{\lambda}}(\mathbf{O}^r))} \mathfrak{D}(\tilde{\alpha}_{j'}^r(1) \sqcap \tilde{\beta}_{j'}^r(1))$;

**6**     $\hat{a}_{jj'} =$
$$
\dfrac{\sum_{r=1}^{R} \dfrac{1}{\mathfrak{D}(h_{\tilde{\lambda}}(\mathbf{O}^r))} \sum_{i=1}^{I_r-1} \mathfrak{D}(\tilde{\alpha}_j^r(i) \sqcap \tilde{a}_{jj'} \sqcap \tilde{b}_{j'}(o_{i+1}^r) \sqcap \tilde{\beta}_{j'}^r(i+1))}{\sum_{r=1}^{R} \dfrac{1}{\mathfrak{D}(h_{\tilde{\lambda}}(\mathbf{O}^r))} \sum_{i=1}^{I_r} \mathfrak{D}(\tilde{\alpha}_j^r(i) \sqcap \tilde{\beta}_j^r(i))};
$$

**7**     $\hat{a}_{jJ} = \dfrac{\sum_{r=1}^{R} \dfrac{1}{\mathfrak{D}(h_{\tilde{\lambda}}(\mathbf{O}^r))} \mathfrak{D}(\tilde{\alpha}_j^r(I) \sqcap \tilde{\beta}_j^r(I))}{\sum_{r=1}^{R} \dfrac{1}{\mathfrak{D}(h_{\tilde{\lambda}}(\mathbf{O}^r))} \sum_{i=1}^{I_r} \mathfrak{D}(\tilde{\alpha}_j^r(i) \sqcap \tilde{\beta}_j^r(i))}$;

**8**     $\hat{\mu}_{jm} \leftarrow \dfrac{\sum_{r=1}^{R} \sum_{i=1}^{I_r} L_{jm}^r(i) o_i^r}{\sum_{r=1}^{R} \sum_{i=1}^{I_r} L_{jm}^r(i)}$;

**9**     $\hat{\Sigma}_{jm} \leftarrow \dfrac{\sum_{r=1}^{R} \sum_{i=1}^{I_r} L_{jm}^r(i)(o_i^r - \hat{\mu}_{jm})(o_i^r - \hat{\mu}_{jm})'}{\sum_{r=1}^{R} \sum_{i=1}^{I_r} L_{jm}^r(i)}$;

**10**     $\hat{w}_{jm} \leftarrow \dfrac{\sum_{r=1}^{R} \sum_{i=1}^{I_r} L_{jm}^r(i)}{\sum_{r=1}^{R} \sum_{i=1}^{I_r} L_j^r(i)}$;

**11** end

**Fig. 5.12** The type-2 fuzzy Baum–Welch algorithm

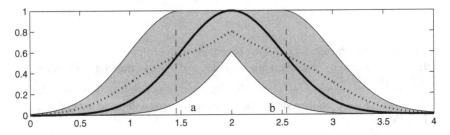

**Fig. 5.13** The *dotted line* and thick *solid line* denote the center of an IT2 MF and the distribution without uncertainty. The intersections *a* and *b* divide the axis into three zones. The *dotted line* is lower than thick *solid line* in the range $[a, b]$ and vice versa, which implies the updating weight induced by the center of the interval is greater than that of the classical Baum–Welch when the input lies outside $[a, b]$

4. Use the final accumulator values to calculate new parameter values to produce a new IT2 FHMM.

5. If the average membership grade of all training observations $\dfrac{\sum_{r=1}^{R}[(h_{\tilde{\lambda}}(\mathbf{O}^r) + \overline{h}_{\tilde{\lambda}}(\mathbf{O}^r))/2]}{R}$ for this iteration is not higher than the value at the previous iteration then stop, otherwise repeat the above steps using the new IT2 FHMM.

## 5.2.5 Type-Reduction and Defuzzification

The type-reduction and defuzzification play important roles in making decisions. For the general T2 FHMM, the output is a T1 FS describing the bounded uncertainty due to randomness and fuzziness in pattern recognition problems. For the IT2 FHMM, the output is an interval set, and the range of the interval set evaluates the bounded uncertainty of both *data* and *model*. The type-2 fuzzy Bayesian decision theory in Sect. 4.1 provides the decision-making rule based on the interval sets. In addition, if we use each T2 FHMM as a rule in the FLSs, we can obtain a defuzzified crisp value from the output interval sets by the KM algorithm as shown in Fig. 3.7.

Based on the T2 fuzzy Bayesian decision theory, we have to rank the output interval sets. We have found it necessary to use the two end points as well as the range of the interval set for classification, so we fuzzify the output interval sets into T1 isosceles triangular fuzzy numbers. The decision rules are as follows.

1. If $\underline{h}_1 < \underline{h}_2$ and $\overline{h}_1 < \overline{h}_2$, then $h_1 < h_2$.
2. If $\underline{h}_1 < \underline{h}_2$ and $\overline{h}_1 > \overline{h}_2$, then $h_1 \approx h_2$. We have to compare them further by the areas of two shaded regions $S_1$ and $S_2$.
3. If $S_1 > S_2$, then $h_1 < h_2$, and vice versa, as illustrated in Fig. 5.14.

We classify the observation sequence to the IT2 FHMM or IT2 FMRF that produces the largest interval $h_{\tilde{\lambda}}(\mathscr{S}^*, \mathbf{O})$ or $U_{\tilde{\lambda}}(\mathscr{S}^*, \mathbf{O})$. Usually, misclassification occurs when output interval sets have a large overlap.

We can also use the IT2 FHMM as each rule in the IT2 FLS. For each class $\tilde{\lambda}_\omega$ with a rule base of $M$ rules, the $l$th rule, $1 \le l \le M$, is

$$R^l : \text{ IF } \mathbf{O} \text{ is IT2 FHMM}^l \text{ THEN } \mathbf{O} \text{ is classified to } \tilde{\lambda}_1 \ (+1) \ [\text{or to } \tilde{\lambda}_2 \ (-1)],$$
$$(5.16)$$

$$R^l : \text{ IF } \mathbf{O} \text{ is IT2 FMRF}^l \text{ THEN } \mathbf{O} \text{ is classified to } \tilde{\lambda}_1 \ (+1) \ [\text{or to } \tilde{\lambda}_2 \ (-1)].$$
$$(5.17)$$

After using KM algorithm in Fig. 3.7 to combine $M$ rules in each class, we obtain an output interval set $\tilde{y}_\omega$ for each class. Based on the classical Bayesian decision theory, we classify the observation $\mathbf{O}$ to $\tilde{\lambda}_\omega$ that has the largest center of $\tilde{y}_\omega$. For simplicity, we can also use the similar system in Fig. 4.5 to perform multi-category classification based on T2 FHMMs.

## 5.2.6 Computational Complexity

The computational complexity of the classical forward-backward algorithm is $O(JI^2)$. Similar to the view that an IT2 FS is a set of embedded T1 FSs, an IT2 FHMM can be considered as a union of embedded HMMs as shown in Fig. 5.15. In this sense the IT2 fuzzy forward-backward algorithm seems to compute two

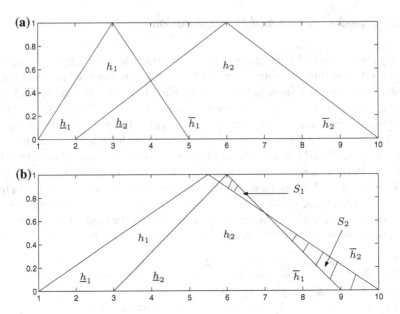

**Fig. 5.14** Two fuzzified IT1 sets $h_1$ and $h_2$ are compared. In **a**, if $\underline{h}_1 < \underline{h}_2$ and $\overline{h}_1 < \overline{h}_2$, then $h_1 < h_2$. In **b**, we have to further compare the two shaded regions $S_1$ and $S_2$. If $S_1 > S_2$, then $h_1 < h_2$, and vice versa

embedded HMMs: the "lower" HMM and the "upper" HMM. Intuitively, the computational complexity of the IT2 FHMM is as twice as that of the HMM. Similar to random uncertainties that flow through the HMM and their effects are evaluated using the mean vectors and variance matrix, fuzzy and random uncertainties flow through the IT2 FHMM, and their effects are evaluated using the the type-reduced and defuzzified output of the IT2 FHMM. If we choose the product $t$-norm in meet operation in the IT2 fuzzy forward-backward algorithm, the bounded sum $t$-conorm in join operation in the IT2 fuzzy forward-backward algorithm, and the maximum $t$-conorm in join operation in the IT2 fuzzy Viterbi algorithm, the IT2 FHMM reduces to the classical HMM when all uncertainties disappear. Therefore, the incorporation of IT2 FS greatly increases the expressive power of the HMM for uncertainty while retaining the tractable learning and recognition procedures.

## 5.3 Speech Recognition

### 5.3.1 Automatic Speech Recognition System

An ASR system includes five components: a speech database with front-end acoustic processing, acoustic models, language models, training algorithm, and recognition

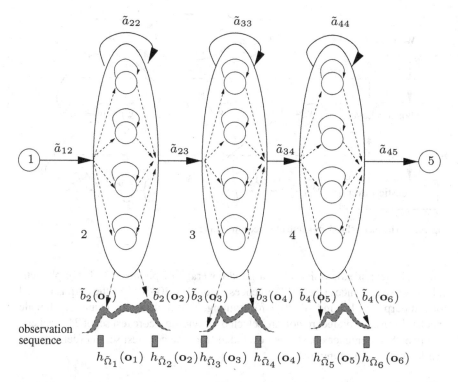

**Fig. 5.15** An example of the left-right IT2 FHMM that is a set of embedded HMMs. Uncertain sequential data are represented by IT2 fuzzy vectors $h_{\tilde{\Omega}}(\mathbf{o}_i)$. The IT2 MF $\tilde{b}_j(\mathbf{o}_i)$ reflects the HMMs uncertainty. The IT1 set $\tilde{a}_{jj'}$ describes the uncertainty in the transition probability

algorithm, in which the acoustic models and the language models are crucial in ASR. The HMM reduces a non-stationary process to a piecewise-stationary process. Phonetic units—phonemes—can be divided into three stationary parts: initial, central and final parts. Thus the HMM is a good acoustic model for phonemes. Figure 5.16 shows the hierarchical structure of HMM-based speech modeling. The within-HMM transitions are determined from the HMM parameters. The between-model transitions are constant. The word-end transitions are determined by the language model with word-level networks.

The language model is used to compute the probability $p(W)$ of a sequence of words, $W = w_1, \ldots, w_L$, and the *bigram* and *trigram* are two widely used models. A useful method to evaluate the impact of the language model on recognition accuracy is *perplexity*, defined as the geometric mean of the number of words that can follow a word after the language model has been applied [2].

TIMIT [3] is a widely used speech database that contains 6,300 utterances produced by 630 speakers from eight major dialect divisions of the United States. For each speaker, there are (1) Two dialect sentences (the "SA" sentences), which meant to expose dialect variants of the speakers; (2) Five phonetically compact sentences

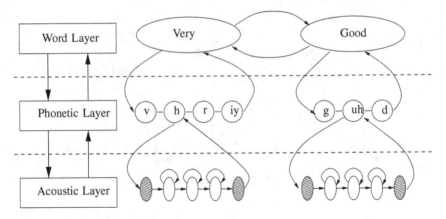

**Fig. 5.16** The hierarchical structure of speech modeling

(the "SX" sentences), which offer a good coverage of phones; (3) Three phonetically diverse sentences (the "SI" sentences). Roughly 20–30 % of the corpus is used
for test purposes, leaving 70–80 % for training. Two male speakers and one female
speaker from each dialect region are selected, providing a core test set of 24 speakers.
The two "SA" sentences have been excluded from the core test set in order to avoid
overlap with the training material.

### 5.3.2 Phoneme Classification

In phoneme classification, we used phoneme boundaries during the test. Besides
the insufficient training data, the noise is a major reason for uncertain parameters
of the statistical model. In this subsection, we carried out extensive experiments on
TIMIT [3] speech database to confirm the effectiveness of T2 FHMMs in the babble
noise.

   In a rough experimental setting, we converted the TIMIT transcription files into
broad-transcription files in which phonemes were broadly grouped as C—consonant,
V—vowel, N—nasal, L—liquid, and S—silence. We selected about 300–500 training
utterances for each phoneme from 10 speakers in the training set. We also selected
about 300–500 utterances for each phoneme from 10 speakers in the core test set
for classification. All these utterances were parameterized using 39 coefficients (one
energy coefficient + 12 cepstrum coefficients + their first and second derivatives).

   To compare the classification performance, for simplicity, we used five labels
and three mixtures in both the HMM and IT2 FHMM (GMMs with uncertain
mean vector), which are adequate to produce good results in this experiment.
We implemented the IT2 fuzzy forward-backward algorithm and the Viterbi

algorithm on MATLAB$^{©}$. The uncertainty factor $k = 3$ and the range of transition, $[\underline{a}_{jj'}, \overline{a}_{jj'}] = [0.98 * a_{jj'}, 1.02 * a_{jj'}]$.

We corrupted test data by white Gaussian noise with different signal-to-noise ratios (SNRs): 5, 10, 15, 20, 25, and 30 dB. Tables 5.1 and 5.2 show the classification results by the HMM and the IT2 FHMM. The average classification rate of the IT2 FHMM is higher than that of the HMM at all SNR levels: 11.9 % higher in 5 dB, 11.9 % higher in 10 dB, 7.2 % dB higher in 15 dB, 5.3 % higher in 20 dB, 2.8 % higher in 25 dB, and 3.1 % higher in 30 dB. We can see that the IT2 FHMM has a comparable classification ability to the HMM in clean speech data. However, when the SNR is low, for example, lower than 20 dB, the IT2 FHMM has a much higher classification rate than the HMM. Figure 5.17 compares the performance of the IT2 FHMM and HMM at different SNR levels. From Fig. 5.17, we can see that the classification rate for the C (consonant) phonemes drops as the SNR rises. This phenomenon may be caused by that the test samples with added noise become increasingly like consonant phonemes. As the experimental results suggest, the proposed IT2 FHMM still outperforms the HMM in this instance.

In a more detailed experimental setting, we broke up 6,300 utterances into phonemes according to the transcriptions, and obtained totally 61 phonemic and phonetic classes including silence and closure intervals of stops. We combined six closure intervals of stops, *bcl, dcl, gcl, pcl, tcl, kcl*, into one category, and combined the "silence" symbols, *pau, epi, h#*, into another category. We deleted eight phonemic classes, *ax-h, b, d, dx, em, eng, nx, zh*, which have less than 500 samples. Thus we obtained totally 46 phonemic categories for classification.

**Table 5.1**  Classification rate (%) of the HMM with different SNRs

| Type of testing data | 5 dB | 10 dB | 15 dB | 20 dB | 25 dB | 30 dB |
|---|---|---|---|---|---|---|
| S | 77.5 | 86.0 | 90.7 | 93.8 | 95.3 | 95.3 |
| C | 94.2 | 90.7 | 89.7 | 85.7 | 85.7 | 85.2 |
| V | 0.9 | 14.4 | 27.5 | 35.5 | 38.3 | 38.2 |
| L | 20.9 | 41.8 | 61.2 | 69.4 | 78.4 | 86.1 |
| N | 0.0 | 7.0 | 21.7 | 45.4 | 63.8 | 76.2 |
| Average | 38.7 | 48.0 | 58.2 | 66.0 | 72.3 | 76.2 |

**Table 5.2**  Classification rate (%) of the IT2 FHMM with different SNRs

| Type of testing data | 5 dB | 10 dB | 15 dB | 20 dB | 25 dB | 30 dB |
|---|---|---|---|---|---|---|
| S | 84.5 | 87.6 | 92.2 | 93.8 | 95.3 | 95.3 |
| C | 96.3 | 94.0 | 92.9 | 90.0 | 90.1 | 88.9 |
| V | 10.7 | 31.8 | 36.3 | 41.1 | 40.5 | 41.0 |
| L | 55.4 | 58.2 | 66.2 | 75.2 | 77.7 | 90.8 |
| N | 6.3 | 28.0 | 39.5 | 56.6 | 71.7 | 80.4 |
| Average | 50.6 | 59.9 | 65.4 | 71.3 | 75.1 | 79.3 |

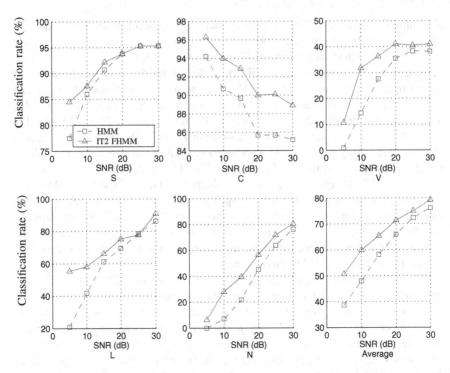

**Fig. 5.17** Classification results of the HMM and IT2 FHMM for five broad phonemes: S (silence), C (consonant), V (vowel), L (liquid), and N (nasal). Average is the average classification rate of S, C, V, L and N. Experimental results show that the IT2 FHMM is more robust to noise than the classical HMM under different signal-to-noise ratios (SNRs)

The speech signal, sampled at 16 KHz mono, was processed in frames of 25 ms with a 15 ms overlapping (rate = 100 Hz). We first pre-emphasized speech frames with an FIR filter (HZ $= 1 - az^{-1}$, $a = 0.97$), and weighted them with a Hamming window to avoid spectral distortions. After pre-processing, we extracted Mel-Frequency Cepstral Coefficents (MFCCs) [2] as the acoustic features. Each acoustic feature vector consists of 12 MFCCs, the energy term, and the corresponding velocity and acceleration derivatives. We did not consider the observation sequences with $T \leq 3$. The dimensionality of acoustic feature vector is 39 for each frame. Thus each observation sequence is a $T \times 39$ matrix. We added the babble noise to all speech data. The babble noise consists of a large number of speakers talking simultaneously. The automatic speech recognition system is very sensitive to the multi-talker, non-stationary babble noise.

To compare the phoneme classification performance in the babble noise, we used five states and three mixtures in the HMM to model phonemes [2]. For each category, we selected 300 training samples from clean data and 200 test samples from data corrupted by the babble noise with signal-to-noise ratio (SNR) 20, 10, 5, 0, −5, and −10 dB. In the 46 phoneme classification, we have to estimate the GLM by the

**Table 5.3** The comparison between HMMs and T2 FHMM-UMs

| SNR | HMM | T2 FHMM-UMs | | |
|---|---|---|---|---|
| | | $k_m = 1$ | $k_m = 2$ | $k_m = 3$ |
| Clean | 54.90 | 57.84 | 58.13 | 56.95 |
| 20 dB | 45.08 | 47.53 | 47.36 | 46.18 |
| 10 dB | 30.66 | 31.64 | 31.70 | 30.47 |
| 5 dB | 22.59 | 23.09 | 23.72 | 22.34 |
| 0 dB | 15.36 | 15.96 | 16.85 | 15.92 |
| −5 dB | 10.01 | 10.36 | 10.71 | 10.36 |
| −10 dB | 5.93 | 6.59 | 6.95 | 6.99 |

**Table 5.4** The comparison between HMMs and T2 FHMM-UVs

| SNR | HMM | T2 FHMM-UVs | | |
|---|---|---|---|---|
| | | $k_v = 0.9$ | $k_v = 0.7$ | $k_v = 0.5$ |
| Clean | 54.90 | 57.02 | 57.10 | 56.99 |
| 20 dB | 45.08 | 46.88 | 47.40 | 47.51 |
| 10 dB | 30.66 | 31.75 | 32.42 | 32.16 |
| 5 dB | 22.59 | 24.16 | 23.79 | 23.45 |
| 0 dB | 15.36 | 16.83 | 16.62 | 16.18 |
| −5 dB | 10.01 | 11.36 | 10.79 | 10.35 |
| −10 dB | 5.93 | 7.01 | 6.50 | 6.39 |

feature vector of 92 length, which is a high dimension leading to bad estimation of the GLM. So, we reduced the dimension 92–46 by PCA [4].

The 53.58 % classification rate of HMMs was reported in [5]. We obtained a similar result 54.90 % in clean test data for HMMs. Table 5.3 compares HMMs and T2 FHMM-UMs for $k_m = 1, 2, 3$. It shows that the classification rate in clean data increases 2.94, 3.23, and 2.05 %, respectively. In particular, when $k_m = 2$, the T2 FHMM-UM performs the best at classifying phonemes corrupted by the babble noise. It outperforms the HMM by increasing the average classification rate 2.28, 1.04, 1.13, 1.49, 0.70, and 1.02 % under 20, 10, 5, 0, −5, and −10 dB, respectively. The classification rate has the std 0.53 for different values of $k_m$, which is consistent with the previous observation that the classification performance is insensitive to different areas of the FOU. Another interesting observation is that when the SNR of the babble noise decreases, the larger FOU in the T2 FHMM-UM obtains slightly better result, for example, when SNR is −10 dB, $k_m = 3$ performs the best. This is also consistent with our intuition that the larger FOU may have more positive effect on the description of the uncertainty, though such an improvement in the classification rate is not that significant.

Table 5.4 compares HMMs and T2 FHMM-UVs for $k_v = 0.9, 0.7, 0.5$. Overall, the T2 FHMM-UV outperforms the HMM consistently. When $k_v = 0.9$, it outperforms the HMM by increasing the average classification rate 1.80, 1.09, 1.57, 1.47,

1.35, and 1.08 % under 20, 10, 5, 0, −5, and −10 dB, respectively. The classification rate is still insensitive to $k_v$ with std 0.32, which means different areas of the FOU may not influence the classification performance significantly. In contrast to the T2 FHMM-UM, the T2 FHMM-UV performs better with the smaller FOU when the SNR of the babble noise decreases.

Although we may choose the arbitrary $k_m$ or $k_v$ for modeling uncertainty with a relatively stable performance, practically we suggest a median value, e.g., $k_m \in$ [1, 2], or $k_v \in$ [0.7, 0.9], according to experimental results.

### 5.3.3 Phoneme Recognition

In phoneme recognition, we evaluated the recognition results by "percent correct" and "accuracy" [2]. We transcribed the 61 phonetic labels defined in TIMIT to 39 labels [3]. For comparison, we used the same number of labels and mixtures in both the HMM and IT2 FHMM (GMMs with uncertain covariance matrix) for each phoneme. We assigned phonemes {b d g dx} four labels, {ih ah uh l r y w m n ng ih dh p t k v hh} six labels, {iy eh uw er ch z f th s sh sil} eight labels, and {ae aa ey ay oy aw ow} 10 labels according to their lengths. We used 32 mixtures in each label. Uncertainty factors $k_1 = 0.95$ and $k_2 = 1.05$. For bench-marking, we used the hidden Markov model toolkit (HTK) to implement the HMM [2]. The training set of the TIMIT database consists of 462 speakers from eight dialect regions. There are total of 3,696 utterances except the "SA" sentences. We used the same training samples for both the HMM and IT2 FHMM, and a bigram language model in this experiment. Finally, we evaluated these two models on the TIMIT core test set [3].

Table 5.5 shows the comparison of the IT2 FHMM with other TIMIT phoneme recognizers. Lee et al. reported the first results of phoneme recognition on the TIMIT database [6]. They selected 48 phonemes to model and used the bigram language model. The features were 12 cepstral coefficients after front-end processing. They could achieve recognition accuracy of 53.27 % in context-independent condition. Young used 48 context-independent left-right HMMs of three labels with diagonal covariance, and obtained 52.7 % recognition accuracy [7]. Glass et al. got 64.1 % recognition accuracy on the core test set by 50 diagonal Gaussian Mixtures [8]. Becchetti et al. used 10 Gaussian mixtures and generated 48 models. They did not use "SA" sentences and reported 62.91 % accuracy on the core test set [9]. The fuzzy GHMM was not applied to phoneme recognition but digit recognition, and had almost the same recognition rate as that of the HMM besides a shorter training time [10]. From Table 5.5, we can see that the IT2 FHMM has at least a comparable recognition rate to the above phoneme recognizers. We also evaluated the IT2 FHMM by "SA" dialect sentences. The uncertainty factors $k_1 = 0.90$ and $k_2 = 1.10$ in this experiment. Table 5.6 shows that the recognition rate of the IT2 FHMM is 5.55 % higher than that of the HMM, which demonstrates that the IT2 FHMM is more robust to large dialect variations in speech data.

**Table 5.5** Comparison with other TIMIT phoneme recognizers (context-independent)

| Model | Correct (%) | Accuracy (%) | Insertion (%) | Substitution (%) | Deletion (%) |
|---|---|---|---|---|---|
| Kai-Fu Lee | 64.07 | 53.27 | 10.79 | 26.22 | 9.72 |
| S. Yonug | 61.70 | 52.70 | – | – | – |
| J. Glass | – | 64.10 | – | – | – |
| C. Becchetti | 67.43 | 62.91 | – | – | – |
| GHMM | – | – | – | – | – |
| HTK | 66.32 | 62.59 | 3.73 | 20.69 | 12.99 |
| IT2 FHMM | 66.60 | 62.94 | 3.66 | 20.89 | 12.51 |

**Table 5.6** Dialect recognition (context-independent)

| Model | Correct (%) | Accuracy (%) | Insertion (%) | Substitution (%) | Deletion (%) |
|---|---|---|---|---|---|
| HTK | 54.17 | 51.39 | 2.78 | 30.56 | 15.28 |
| IT2 FHMM | 56.94 | 56.94 | 0.00 | 30.56 | 12.50 |

Phoneme recognition was conducted without boundary information, which means recognition was performed in word layer illustrated in Fig. 5.16. The between-model transitions are governed by the language model that determines the boundary between phonemes in a word. The language model is the main reason why the IT2 FHMM dose not have much better recognition rate as shown in Table 5.5. However, from Table 5.6 we can see that the IT2 FHMM still outperforms the classical HMM in dialect recognition.

## 5.4 Summary

Theoretically, HMMs can model any sequential data by a simple mechanism: the Markov property in the descriptive level and the GMMs in the generative level. We assume the observations generated by hidden labels are i.i.d., which can be satisfied by proper feature extraction methods. The relationship of observations can be further reflected by transition probabilities between labels. Such a hierarchical mechanism provides an advantage to separately describe the statistical information of observations and the structural information of labels within a unified framework. This is one of the reasons why HMMs gain much attention for sequential data modeling in the past forty years [1, 11–13].

As we see, HMMs have the simplest causal neighborhood system, that is, the label at time $i$ only depends on the label at time $i - 1$. We further impose left-right constraints on the transition probability and use GMMs as the generating distributions. Therefore, in HMMs, both the descriptive and the generative models are so simple that leads to efficient algorithms, such as the forward-backward, Viterbi and Baum-Welch, for finding the best labeling configuration and learning from data [1].

However, in his seminal paper [1], Rabiner mentioned three inherent limitations of HMMs for practical use:

1. The statistical independence assumption that successive observations are independent so that the joint probability of a sequence of observations can be written as a product of probabilities of individual observations;
2. The assumption that the observation distributions can be well represented as a mixture of Gaussian models;
3. The first-order Markov assumption that the probability of being at time $i$ depends only at the time $i - 1$.

In speech recognition, the statistical independence assumption may be satisfied by the Mel-frequency cepstral coefficients (MFCC) front-end processing so that each acoustic vector is uncorrelated with its neighbors [2, 14]. As far as the first-order Markov assumption is concerned, some schemes have explored a longer history of the process, but high-computational complexity makes them intractable [15]. *Kullback-Leibler divergence* is a practical method to dynamically determine the statistical dependent sites, such as variable length Markov models (VLMMs) [16] and statistical character modeling [17]. The most efforts have been put into relaxing the second limitation, i.e., enhancing the "expressive power" of the HMM [9, 11, 13, 18]. Some researchers have used artificial neural networks in each state as observation distributions [19–21]. Hidden semi-Markov models (HSMMs) can generate a sequence of observations at each site [22]. Factorial HMMs can represent the combination of multiple signals using distinct Markov chains [23], and coupled HMMs can model audio-visual signals simultaneously in the noisy environment [24]. Fuzzy Markov chains model the uncertainty of transition probability using fuzzy numbers [25]. The linguistic HMM can process fuzzy feature vectors [26]. Generalized HMMs can relax the statistical independence limitation and additivity constraint in the probability measure by fuzzy measures and fuzzy integral [10, 27, 28]. Type-2 fuzzy HMMs can handle two kinds of uncertainties, namely, randomness and fuzziness, simultaneously within a unified framework [29].

Another major problem is how to determine the topology (the number of labels and mixtures) of HMMs based on characteristics of the problem domain [13]. HMMs assume that sequential data are piecewise stationary [1, 14]. At each stationary piece, HMMs use a hidden label to represent the statistical variations of underlying observations, so the number of labels is fewer than the number of observations, i.e., $J < I$. Generally, more mixture components in GMMs can describe such variations more accurately given sufficient training samples [13]. Therefore, we prefer more labels and mixtures for the better modeling ability. But because of limited training samples to train parameters of GMMs, we often have to make a tradeoff between performance and complexity for practical problems. To compromise this contradiction, Kwong used genetic algorithms to optimize the topology and parameters of HMMs [30]. Usually, we have to design the topology based on prior knowledge of the structure in sequential data, such as phoneme, protein, and nucleic acid. Phonemes can be divided into three stationary parts: initial, central,s and final parts. So a five label HMM is suitable for phoneme modeling. A modified hidden Markov model, TMHMM, can predict transmembrane helices in protein sequences successfully because its topology maps closely to the biological system [31]. Bagos introduced how to incorporate topological information of transmembrane proteins in HMMs [32].

The implementation of HMMs for speech recognition by C and C++ can be found in [2, 9]. Other applications of HMMs are shown in [12], such as speech recognition [1], handwriting recognition [33, 34], gesture recognition [34–36], human motion [16, 37], gene modeling [38], and face detection and recognition [39].

Indeed, the linguistic hidden Markov model (LHMM) [26] and linguistic fuzzy c-means [40] can also compute fuzzy vectors. However, the proposed T2 FHMM is different from LHMM in three important aspects: (1) We represent uncertainty of data as a sequence of T2 fuzzy vectors rather than T1 fuzzy vectors; (2) We use T2 FS to model HMM's fuzziness rather than a union of different level interval HMMs used by LHMM; (3) We extend the forward-backward algorithm and Viterbi algorithm using T2 FS operations rather than the extension principle and decomposition theorem. Though both the T2 FHMM and LHMM use interval arithmetic for computation in practice, the T2 FHMM is a T2 and LHMM is a T1 fuzzy extension of the classical HMM. Generalized hidden Markov models (GHMMs) [27, 28] relax the statistical independence limitation and additivity constraint in the probability measure by fuzzy measures and fuzzy integral, whereas the T2 FS, with its solid theoretical background and practical success in signal processing, is probably to shed more light on building a theoretically well-founded framework that enhances the HMM's expressive power for modeling uncertain sequential data.

In Sect. 3.5, we use the T2 fuzzy similarity measure between the input nonsingle-ton fuzzified data $\tilde{x}$ and the T2 fuzzy class model $\tilde{\lambda}$. From this perspective, the T2 fuzzy forward-backward, Viterbi, and relaxation labeling algorithms are particular for T2 similarity measure between input T2 fuzzy vectors and T2 FHMMs. In the IT2 FHMMs we only measures the similarity between two embedded T1 FSs, the lower and upper MFs in Eq. 3.55. Hence, the computational complexity of all extended algorithms by IT2 FSs operations is tractable.

The IT2 FHMM provides additional flexibility for classification. We can use either left-end, central, or right-end point of the interval set to classify data. Without loss of generality, we propose to rank the interval sets by triangular T1 fuzzy numbers. Currently, we have not investigated the T2 FLSs for real-world pattern recognition problems, but we propose to use IT2 FHMMs as the rule base in T2 FLSs. Through the KM algorithm, we can obtain the centroid of the output interval sets for classification.

This chapter has shown the IT2 FHMM that can effectively handle both randomness and fuzziness. The output of the IT2 FHMM is an uncertain T1 FS rather than a crisp scalar in the classical HMM. In this way, both random and fuzzy uncertainties can be accounted for in a unified framework. In order to realize the IT2 FHMM, we model each hidden label by a IT2 MF $\tilde{b}_j(\mathbf{o}_i)$ with two kinds of forms: GMMs with uncertain mean vector and covariance matrix. Meanwhile, the transition probability is modeled by a T1 fuzzy number. Experimental results have shown that the IT2

FHMM has a higher classification rate than the classical HMM in clean speech data, and is more robust to speech signals with noise and large dialect variations.

# References

1. Rabiner, L.R.: A tutorial on hidden Markov models and selected applications in speech recognition. Proc. IEEE **77**, 257–286 (1989)
2. Young, S., Evermann, G., Kershaw, D., Moore, G., Odell, J., Ollason, D., Povey, D., Valtchev, V., Woodlands, P.: The HTK Book for HTK Version 3.2. Cambridge University Engineering Department, Cambridge, UK (2002)
3. Garofolo, J.S., Lamel, L.F., Fisher, W.M., Fiscus, J.G., Pallett, D.S., Dahlgren, N.L.: DRAPA TIMIT acoustic-phonetic continuous speech corpus CD-ROM. NISTIR 4930 (1992)
4. Duda, R.O., Hart, P.E., Stork, D.G.: Pattern Classification, 2nd edn. Wiley, New York (2001)
5. Golowich, S.E., Sun, D.X.: A support vector/hidden Markov model approach to phoneme recognition. ASA Proceedings of the Statistical Computing Section pp. 125–130 (1998)
6. Lee, K.F., Hon, H.W.: Speaker-independent phone recognition using hidden Markov models. IEEE Trans. Acoust. Speech Signal Process. **37**(11), 1641–1648 (1989)
7. Young, S.: The general use of tying in phoneme-based HMM speech recognizers. In: Proceedings of the IEEE ICASSP pp. 569–572 (1992)
8. Glass, J., Chang, J., McCandless, M.: A probabilistic framework for feature based speech recognition. Proceedings of the IEEE ICASSP pp. 2277–2280 (1996)
9. Becchetti, C., Ricotti, L.P.: Speech Recognition Theory and C++ Implementation. Wiley, New York (1999)
10. Chevalier, S., Kaynak, M.N., Cheok, A.D., Sengupta, K.: Use of a novel non-linear generalized fuzzy hidden Markov model for speech recognition. Int. J. Control Intell. Syst. Spec. Issue Non-Linear Speech Recognit. **30**(2), 68–82 (2002)
11. Bengio, Y.: Markovian models for sequential data (1999). http://www.icsi.berkeley.edu/~jagota/NCS
12. Cappé, O.: Ten years of HMMs (2001). http://www.tsi.enst.fr/~cappe/docs/hmmbib.html
13. Bilmes, J.: What HMMs can do. UWEE Tech. Rep. UWEETR-2002-2003, Department of EE, University of Washington, Seattle (2002)
14. Young, S.: A review of large-vocabulary continuous-speech recognition. IEEE Signal Process. Mag. **13**(5), 45–56 (1996)
15. Saul, L., Jordan, M.I.: Exploiting tractable substructures in intractable networks. Adv. Neural Inf. Process. Syst. **8**, 486–492 (1996)
16. Galata, A., Johnson, N., Hogg, D.: Learning behavior models of human activities. British Machine Vision Conference pp. 12–22 (1999)
17. Kim, I.J., Kim, J.H.: Statistical character structure modeling and its application to handwriting chinese character recognition. IEEE Trans. Pattern Anal. Mach. Intell. **25**(11), 1422–1436 (2003)
18. Nakagawa, S.: A survey on automatic speech recognition. IEICE Trans. Inf. Syst. **E85-D**(3), 465–486 (2002)
19. Bengio, Y.: Neural Networks for Speech and Sequence Recognition. International Thomson Computer Press, London (1996)
20. Morgan, H., Bourlard, H.: Continuous speech recognition using multilayer perceptrons with hidden Markov models. Proc. IEEE ICASSP **77**, 413–416 (1990)
21. Robinson, A.J.: An application of recurrent nets to phone probability estimation. IEEE Trans. Neural Netw. **5**(2), 298–305 (1994)
22. Murphy, K.: Hidden semi-Markov models (HSMMs) (2002). www.ai.mit.edu/~murphyk
23. Ghahramani, Z., Jordan, M.I.: Factorial hidden Markov models. In: Proceedings of the Conference Advances in Neural Information Processing Systems, NIPS 8, 472–478 (1995)

24. Nefian, A.V., Liang, L.H., Liu, X.X., Pi, X., Mao, C., Murphy, K.: A coupled HMM for audio-visual speech recognition. Proc. IEEE ICASSP **2**, 2013–2016 (2002)
25. Buckley, J.J.: Fuzzy Probabilities : New Approach and Applications. Physica-Verlag, New York (2003)
26. Popescu, M., Keller, J., Gader, P.: Linguistic hidden Markov models. Proceedings of the FUZZ-IEEE pp. 796–801 (2003)
27. Mohamed, M.A., Gader, P.: Generalized hidden Markov models - part I: Theoretical frameworks. IEEE Trans. Fuzzy Syst. **8**(1), 67–81 (2000)
28. Mohamed, M.A., Gader, P.: Generalized hidden Markov models - part II: application to handwritten word recognition. IEEE Trans. Fuzzy Syst. **8**(1), 82–94 (2000)
29. Zeng, J., Liu, Z.Q.: Type-2 fuzzy hidden Markov models and their application to speech recognition. IEEE Trans. Fuzzy Syst. **14**(3), 454–467 (2006)
30. Kwong, S., Chau, C.W., Man, K.F., Tang, K.S.: Optimization of HMM topology and its model parameters by genetic algorithms. Pattern Recognit. **34**(2), 509–522 (2001)
31. Krogh, A., Larsson, B., von Heijne, G., Sonnhammer, E.L.L.: Predicting transmembrane protein topology with a hidden Markov model: application to complete genomes. J. Math. Imaging Vis. **305**(3), 567–580 (2001)
32. Bagos, P.G., Liakopoulos, T.D., Hamodrakas, S.J.: Algorithms for incorporating prior topological information in HMMs: application to transmembrane proteins. BMC Bioinform. **7**, 189 (2006)
33. Liu, Z.Q., Cai, J., Buse, R.: Handwriting Recognition : Soft Computing and Probabilistic Approaches. Springer, Berlin (2003)
34. Bunke, H., Caelli, T. (eds.): Hidden Markov Models : Applications in Computer Vision. World Scientific, River Edge (2001)
35. Wilson, A.D., Bobick, A.F.: Parametric hidden Markov models for gesture recognition. IEEE Trans. Pattern Anal. Mach. Intell. **21**(9), 884–900 (1999)
36. Lee, D.D., Seung, H.S.: Learning the parts of objects by non-negative matrix factorization. Nature **401**, 788–791 (1999)
37. Niwase, N., Yamagishi, J., Kobayashi, T.: Human walking motion synthesis with desired pace and stride length based on HSMM. IEICE Trans. Inf. Syst. **E88–D**(11), 2492–2499 (2005)
38. Koski, T.: Hidden Markov Models for Bioinformatics. Kluwer Academic Publishers, London (2001)
39. Li, S.Z., Jain, A.K. (eds.): Handbook of Face Recognition. Springer, New York (2005)
40. Auephanwiriyakul, S., Keller, J.M.: Analysis and efficient implementation of a linguistic fuzzy c-means. IEEE Trans. Fuzzy Syst. **10**(5), 563–582 (2002)

# Chapter 6
# Type-2 Fuzzy Markov Random Fields

**Abstract** This chapter integrates type-2 fuzzy sets (T2 FSs) with Markov random fields (MRFs) referred to as T2 FMRFs, which may handle both fuzziness and randomness in the structural pattern representation. On the one hand, the T2 membership function (MF) has a three-dimensional structure in which the primary MF describes randomness, and the secondary MF evaluates the fuzziness of the primary MF. On the other hand, MRFs can represent patterns statistical-structurally in terms of neighborhood system $\partial i$ and clique potentials $V_c$, and thus have been widely applied to image analysis and computer vision. In the proposed T2 FMRFs, we define the same neighborhood system as that in classical MRFs. To describe uncertain structural information in patterns, we derive the fuzzy likelihood clique potentials from T2 fuzzy Gaussian mixture models (T2 FGMMs). The fuzzy prior clique potentials are penalties for the mismatched structures based on prior knowledge. Because Chinese characters have hierarchical structures, we use T2 FMRFs to model character structures in the handwritten Chinese character recognition (HCCR) system. The overall recognition rate is 99.07 %, which confirms the effectiveness of T2 FMRFs for statistical character structure modeling.

## 6.1 Markov Random Fields

Compared with HMMs, MRFs have a more complex neighborhood system at two-dimensional sites. We prefer the noncausal neighborhood system $\mathcal{N}$ because it is difficult to define "cause-and-effect" relations for two-dimensional spatial sites. We assume $\lambda$ is the set of parameters defining an MRF. Therefore, unlike HMMs, we cannot use simple dynamic programming to infer $P(\mathbf{O}|\lambda)$ in Eq. (2.8). Because finding the global best labeling configuration $\mathcal{S}^*$ is an intractable combinatorial problem, we have to approximate the global solution by the local labeling configuration. Such algorithm as relaxation labeling can achieve this goal with polynomial complexity.

Indeed, complex neighborhood systems make MRFs more powerful to represent the complex relationships among image pixels and regions, so that more complex structural information can be accounted for. Unlike HMMs' transition probability, evaluating the joint probability $P(\mathcal{S})$ is more difficult in MRFs due to the

© Tsinghua University Press, Beijing and Springer-Verlag Berlin Heidelberg 2015     85
J. Zeng and Z.-Q. Liu, *Type-2 Fuzzy Graphical Models for Pattern Recognition*,
Studies in Computational Intelligence 591, DOI 10.1007/978-3-662-44690-4_6

combinatorial problem. To overcome the problem, we always seek to evaluate the corresponding energy function according to the equivalence between MRFs and GRFs (Gibbs random fields), in which the energy function is a sum of local clique potentials over all possible cliques. A clique $c$ is defined by a subset of sites that are all pairwise neighbors in $\mathcal{N}$. It may consist either a single-site $\mathscr{C}_1 = \{i\}$, or of a pair of neighboring sites $\mathscr{C}_2 = \{(i, i')\}$, or of a triple of neighboring sites $\mathscr{C}_3 = \{(i, i', i'')\}$. High-order cliques can also represent high-order relationships of labels. For simplicity, we only consider $\mathscr{C}_1$ and $\mathscr{C}_2$ here. We assign costs $V_c$ to different cliques to encourage or penalize different local interactions among labels at neighboring sites. The Hammersley-Clifford theorem establishes the equivalence between MRFs and GRFs, so the likelihood function in Eq. (2.12) can be written as

$$p(\mathbf{O}|\mathscr{S}, \lambda) = Z^{-1} e^{-U(\mathbf{O}|\mathscr{S}, \lambda)}, \qquad (6.1)$$

and the prior probability can be written as

$$P(\mathscr{S}|\lambda) = Z^{-1} e^{-U(\mathscr{S}|\lambda)}, \qquad (6.2)$$

where the likelihood energy

$$U(\mathbf{O}|\mathscr{S}, \lambda) = \sum_{c \in \mathscr{C}_1, \mathscr{C}_2} V_c(\mathbf{O}|\mathscr{S}, \lambda), \qquad (6.3)$$

and the prior energy

$$U(\mathscr{S}|\lambda) = \sum_{c \in \mathscr{C}_1, \mathscr{C}_2} V_c(\mathscr{S}|\lambda). \qquad (6.4)$$

The partition function $Z^{-1}$ is the sum of energy of all possible configurations, which is an ignorable constant in terms of $\mathscr{S}$. So from Eq. (2.12), the posterior energy function

$$U(\mathscr{S}|\mathbf{O}, \lambda) \propto U(\mathbf{O}|\mathscr{S}, \lambda) + U(\mathscr{S}|\lambda) = \sum_{c \in \mathscr{C}_1, \mathscr{C}_2} (\underbrace{V_c(\mathbf{O}|\mathscr{S}, \lambda)}_{\text{statistical information}} + \underbrace{V_c(\mathscr{S}|\lambda)}_{\text{structural information}}), \qquad (6.5)$$

where likelihood clique potentials $V_c(\mathbf{O}|\mathscr{S}, \lambda)$ and prior clique potentials $V_c(\mathbf{O}|\mathscr{S}, \lambda)$ reflect the statistical information as well as the local structural information of patterns. From the Bayesian decision rule (2.13), maximizing the posterior probability is equivalent to minimize the posterior energy. Figure 6.1 shows the MRF to represent both statistical and structural information of patterns.

Up till now, we have converted a global optimization problem into a local one by neighborhood systems and clique potentials. The problem left is how to design proper neighborhood systems, likelihood clique potentials, and prior clique potentials

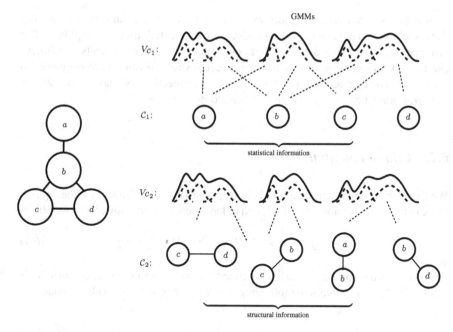

**Fig. 6.1** An MRF has four labels $a$, $b$, $c$, and $d$ with the link for neighboring labels. The likelihood clique potentials are based on GMMs. The single-site clique potential $V_{\mathscr{C}_1}$ describes the statistical information, and the pair-site clique potential $V_{\mathscr{C}_2}$ describes the structural information statistically

to represent the patterns for real-world problems. We continue to use GMMs as generating distributions of the hidden labels. Then, we can derive the likelihood clique potentials from GMMs. We will see that the only difference between MRFs and HMMs lies in the descriptive level where MRFs have a more complex neighborhood system to represent more complex structural information at two-dimensional sites.

We will consider the design of neighborhood systems and clique potentials in turn.

### 6.1.1 The Neighborhood System

We can design neighborhood systems by distance between sites, such as Euclidean distance. The set of neighbors of $i$ is defined as the set of sites within a radius of $\sqrt{r}$ from $i$, that is,

$$\mathcal{N}_i = \{i' \in \mathcal{S} \,|\, [\text{distance}(\text{site}_{i'}, \text{site}_i)]^2 \leqslant r, i' \neq i\}, \tag{6.6}$$

where the sites can be either regular or irregular. In general, the neighborhood system at irregular sites may have varying shapes and sizes.

Another way to design neighborhood system is based on connection of irregular sites, such as image regions. If two regions are connected, they are neighbors. For instance, we can design the neighborhood system for stroke segments of Chinese character by connection, in which two connected stroke segments are neighbors to each other. The neighborhood system defined by connection also has varying shapes and sizes but usually simpler than that defined by distance.

### 6.1.2 Clique Potentials

We first derive single-site likelihood clique potentials from GMMs. By the equivalence between MRFs and GRFs (6.1) and independence assumption (2.14), we obtain

$$p(\mathbf{o}_i|s_i = j, \lambda) = Z^{-1} e^{-V_{\mathscr{C}_1}(\mathbf{o}_i|s_i=j,\lambda)}, \quad 1 \leq j \leq J, \tag{6.7}$$

where $Z^{-1}$ is the constant partition function that can be ignored. Because $p(\mathbf{o}_i|s_i = j, \lambda)$ is a GMM, we obtain the following single-site clique likelihood potentials

$$V_{\mathscr{C}_1}(\mathbf{o}_i|s_i = j, \lambda) = -\log\left[\sum_{m=1}^{M_s} w_{jm} G(\mathbf{o}_i; \boldsymbol{\mu}_{jm}, \boldsymbol{\Sigma}_{jm})\right]. \tag{6.8}$$

Similarly, we can derive pair-site likelihood clique potentials from

$$p(\mathbf{o}_i, \mathbf{o}_{i'}|s_i = j, s_{i'} = j', \lambda) = Z^{-1} e^{-V_{\mathscr{C}_2}(\mathbf{o}_i,\mathbf{o}_{i'}|s_i=j,s_{i'}=j',\lambda)]}, \quad 1 \leq j, j' \leq J, \tag{6.9}$$

where the constant partition function $Z^{-1}$ can be ignored. Thus, the pair-site likelihood clique potentials are

$$V_{\mathscr{C}_2}(\mathbf{o}_i, \mathbf{o}_{i'}|s_i = j, s_{i'} = j', \lambda) = -\log\left[\sum_{m=1}^{M_s} w_{jj'm} G(\mathbf{o}_{ii'}; \boldsymbol{\mu}_{jj'm}, \boldsymbol{\Sigma}_{jj'm})\right], \tag{6.10}$$

where $\mathbf{o}_{ii'}$ is the binary feature, such as the difference or ratio between $\mathbf{o}_{i'}$ and $\mathbf{o}_i$.

The prior clique potentials reflect the local structure in the labeling space by encouraging the connected labels or penalizing the disconnected labels at neighboring sites. We use the conditional probability, $a_{jj'} = P(s_{i'} = j'|s_i = j)$, for this local labeling constraint. If the label $j$ allows its neighboring label $j'$, we say that the label $j$ is connected with $j'$ denoted by $a_{jj'} = 1$, and otherwise $a_{jj'} = 0$,

$$a_{jj'} = \begin{cases} 1, & j \text{ is connected with } j', \\ 0, & j \text{ is disconnected with } j'. \end{cases} \tag{6.11}$$

The concept of connection is similar to the "edge" in graphical models, where two nodes have a certain relationship represented by an edge. If a label is not assigned to any sites denoted by $j = 0$, a penalty is imposed because such a labeling configuration does not satisfy the prior structure in the labeling space; that is

$$V_{\mathscr{C}_1}(j|\lambda) = \begin{cases} 0, & \text{if } j \neq 0, \\ \sum_{j'=1}^{J} (1 + a_{jj'})v_1, & \text{if } j = 0, \end{cases} \tag{6.12}$$

where $v_1 > 0$ is a penalty. To keep the balance with the single-site likelihood clique potential, we define $v_1$ as

$$v_1 = \min_{i=1}^{I} V_{\mathscr{C}_1}(\mathbf{o}_i|s_i = j, \lambda), \tag{6.13}$$

which is the minimum value of the single-site likelihood clique potential of label $j$ for all observations. By the pair-site prior clique potential, we encourage the connected labels and penalize the disconnected labels at neighboring sites,

$$V_{\mathscr{C}_2}(j, j'|\lambda) = -a_{jj'}v_2, \tag{6.14}$$

where $v_2$ is a positive constant that encourages the labeling configuration that is consistent with the prior structure.

### 6.1.3 Relaxation Labeling

Given observations $\mathbf{O}$ and an MRF $\lambda$, the relaxation labeling (RL) is a class of parallel iterative numerical procedures with polynomial complexity to find the best labeling, $\mathscr{S}^* = \{s_1^*, \ldots, s_I^*\}$, and the minimum of the posterior energy $U(\mathscr{S}^*|\mathbf{O}, \lambda)$. In the continuous RL, the labeling assignment is defined as

$$\mathscr{F} = \left\{ f_j(i) \in [0, 1]|1 \leq i \leq I, \sum_{j=1}^{J} f_j(i) = 1 \right\}, \tag{6.15}$$

where $f_j(i)$ is the strength that the label $s_i = j$ is assigned to $\mathbf{o}_i$. We understand this strength as the posterior probability $p(s_i = j|\mathbf{o}_i, \lambda)$, and use this probability as updating weight in the Baum-Welch algorithm.

In practice, we convert the minimization of the posterior energy into the maximization of a corresponding gain function, which is the sum of compatibility functions. The RL compatibility functions can be defined by the following clique potentials,

$$K_j(i) = \text{CONST}_1 - V_{\mathscr{C}_1}(\mathbf{o}_i | s_i = j, \lambda) - V_{\mathscr{C}_1}(j | \lambda), \tag{6.16}$$

and

$$K_{j,j'}(i, i') = \text{CONST}_2 - V_{\mathscr{C}_2}(\mathbf{o}_i, \mathbf{o}_{i'} | s_i = j, s_{i'} = j', \lambda) - V_{\mathscr{C}_2}(j, j' | \lambda). \tag{6.17}$$

The constant $\text{CONST}_1$ and $\text{CONST}_2$ satisfy that all the compatibility functions are non-negative. The gain function with the labeling assignment can now be written as

$$g(\mathscr{S} | \mathbf{O}, \lambda) = \sum_{i=1}^{I} \sum_{j=1}^{J} \left[ K_j(i) f_j(i) + \sum_{i' \in \mathscr{N}_i} \sum_{j'=1}^{J} K_{j,j'}(i, i') f_j(i) f_{j'}(i') \right]. \tag{6.18}$$

We can see that if the compatibility functions are replaced by transition probability $a_{jj'}$ and generating distributions $b_j(\mathbf{o}_i)$, the gain function is actually the sum of forward or backward variables. We can also design the gain function corresponding to the Viterbi algorithm that replace the $\sum$ operator with max operator,

$$g(\mathscr{S} | \mathbf{O}, \lambda) = \sum_{i=1}^{I} \sum_{j=1}^{J} \left[ K_j(i) f_j(i) + \sum_{i' \in \mathscr{N}_i} \max_{j'=1}^{J} K_{j,j'}(i, i') f_j(i) f_{j'}(i') \right]. \tag{6.19}$$

To maximize $g(\mathbf{O}, \mathscr{S} | \lambda)$, the RL iteratively updates $f_j(i)$ by the gradient

$$q_j(i) = \frac{\partial g}{\partial f_j(i)} = K_j(i) + \sum_{i' \in \mathscr{N}_i} \sum_{j'=1}^{J} K_{j,j'}(i, i') f_{j'}(i'), \tag{6.20}$$

or

$$q_j(i) = \frac{\partial g}{\partial f_j(i)} = K_j(i) + \sum_{i' \in \mathscr{N}_i} \max_{j'=1}^{J} K_{j,j'}(i, i') f_{j'}(i'). \tag{6.21}$$

If $f_j^t(i)$ is the ambiguous relaxation labeling after $t$ iterations, the $f_j^{t+1}(i)$ is updated using the following fix-point iteration if $q_j^t(i) \geq 0$,

$$f_j^{t+1}(i) = \frac{f_j^t(i) q_j^t(i)}{\sum_{j=1}^{J} f_j^t(i) q_j^t(i)}. \tag{6.22}$$

Two factors affect the solution of RL: the initial labeling assignment and the compatibility function. Because the MRF compatibility function contains both prior and likelihood information, the RL does not depend on the initial labeling that much.

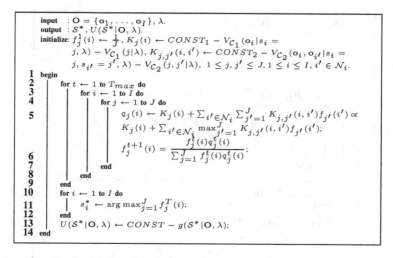

**Fig. 6.2** The relaxation labeling algorithm

After several iterations, the labeling assignment will be unambiguous. The RL terminates if the number of iterations reach a fixed number $T_{\max}$, and we use the winner-take-all strategy,

$$s_i^* = \arg\max_{j=1}^{J} f_j(i). \tag{6.23}$$

The winner-take-all relaxation labeling algorithm is given in Fig. 6.2. The computational complexity depends on the definition of neighborhood system. For the simplest case, if each site has only one neighboring site, the RL has $O(J^2 I)$ complexity. For the most complicated case, if each site has all other sites as neighbors, the RL has $O(I^2 J + J^2 I)$ complexity.

Analogously to HMMs, we can initialize the MRF automatically, and then use the Baum-Welch algorithm to refine $\mu$ and $\Sigma$ in GMMs with the updating weight $f_j(i)$ by relaxation labeling as shown in Figs. 6.3 and 6.4. Note that the only difference is to estimate parameters in the pair-site likelihood clique potential. We ca use a standard observation set, $\mathbf{O}^* = \{\mathbf{o}_1^*, \ldots, \mathbf{o}_I^*\}$, to initialize the MRF, where $J = I$, $\mu_j = \mathbf{o}_i^*$, $\mu_{jj'} = \mathbf{o}_{ii'}^*$, and $a_{jj'} = 1$ if site $i'$ is the neighbor of site $i$. The initial covariance matrix $\Sigma_j$ and $\Sigma_{jj'}$ can be set as a diagonal matrix $diag(\sigma_1^2, \ldots, \sigma_d^2)$, where $\sigma_i = \text{CONST}$. We then update all parameters from training samples.

## 6.2 Type-2 Fuzzy Markov Random Fields

The T2 FMRFs are a class of MRFs, because they satisfy the local labeling constraints of MRFs at neighboring sites in terms of T2 MFs,

$$
\begin{array}{ll}
\textbf{input} & : \mathbf{O}^r = \{o_1^r, \ldots, o_I^r\}, 1 \le r \le R, \mathbf{O}^* = \\
& \quad \{o_1^*, \ldots, o_I^*\}, M_s. \\
\textbf{output} & : \hat{\lambda} = \{\hat{a}_{jj'}, \hat{\mu}_{jm}, \hat{\Sigma}_{jm}, \hat{w}_{jm}, \hat{\mu}_{jj'm}, \hat{\Sigma}_{jj'm}, \hat{w}_{jj'm}\}. \\
\textbf{initialize:} & \Theta_{jm}^r(i), \Theta_{jj'm}^r(ii') \leftarrow 0.
\end{array}
$$

1 **begin**

2    $J, \mu_j, \mu_{jj'}, \Sigma_j, \Sigma_{jj'}, a_{jj'} \leftarrow \mathbf{O}^* = \{o_1^*, \ldots, o_I^*\};$

3    **for** $r \leftarrow 1$ **to** $R$ **do**

4      $U(\mathcal{S}^*|\mathbf{O}^r, \lambda), \mathcal{S}^* \leftarrow \texttt{relaxationlabeling}(\mathbf{O}^r, \lambda);$

5      **if** $s_i^r = j, s_{i'}^r = j'$ **then**

6        $\Theta_j^r(i), \Theta_{jj'}^r(i) \leftarrow 1;$

7      **end**

8      $A_{jj'} \leftarrow$ total number of neighboring observations labeled with $j$ and $j'$;

9    **end**

10    $\Theta_{jm}^r(i), \Theta_{jj'm}^r(i) \leftarrow \texttt{kmeans}(\mathbf{O}^r, \Theta_j^r(i), \Theta_{jj'}^r(i), M_s), \forall r;$

11    $\hat{a}_{jj'} \leftarrow \dfrac{A_{jj'}}{R};$

12    $\hat{\mu}_{jm} \leftarrow \dfrac{\sum_{r=1}^R \sum_{i=1}^{I_r} \Theta_{jm}^r(i) o_i^r}{\sum_{r=1}^R \sum_{i=1}^{I_r} \Theta_{jm}^r(i)};$

13    $\hat{\mu}_{jj'm} \leftarrow \dfrac{\sum_{r=1}^R \sum_{i=1}^{I_r} \Theta_{jj'm}^r(i) o_{ii'}^r}{\sum_{r=1}^R \sum_{i=1}^{I_r} \Theta_{jj'm}^r(i)};$

14    $\hat{\Sigma}_{jm} \leftarrow \dfrac{\sum_{r=1}^R \sum_{i=1}^{I_r} \Theta_{jm}^r(i)(o_i^r - \hat{\mu}_{jm})(o_i^r - \hat{\mu}_{jm})'}{\sum_{r=1}^R \sum_{i=1}^{I_r} \Theta_{jm}^r(i)};$

15    $\hat{\Sigma}_{jj'm} \leftarrow \dfrac{\sum_{r=1}^R \sum_{i=1}^{I_r} \Theta_{jj'm}^r(i)(o_{ii'}^r - \hat{\mu}_{jj'm})(o_{ii'}^r - \hat{\mu}_{jj'm})'}{\sum_{r=1}^R \sum_{i=1}^{I_r} \Theta_{jj'm}^r(i)};$

16    $\hat{w}_{jm} \leftarrow \dfrac{\sum_{r=1}^R \sum_{i=1}^{I_r} \Theta_{jm}^r(i)}{\sum_{r=1}^R \sum_{i=1}^{I_r} \sum_{m=1}^{M_s} \Theta_{jm}^r(i)};$

17    $\hat{w}_{jj'm} \leftarrow \dfrac{\sum_{r=1}^R \sum_{i=1}^{I_r} \Theta_{jj'm}^r(i)}{\sum_{r=1}^R \sum_{i=1}^{I_r} \sum_{m=1}^{M_s} \Theta_{jj'm}^r(i)};$

18 **end**

**Fig. 6.3** The initialization algorithm

$$
h_{\tilde{\lambda}}(\tilde{\mathscr{S}}) > 0, \forall \tilde{\mathscr{S}}, \tag{6.24}
$$

$$
h_{\tilde{\lambda}}(\tilde{s}_i | \tilde{\mathscr{S}}_{\{/i\}}) = h_{\tilde{\lambda}}(\tilde{s}_i | \tilde{\mathscr{N}}_i), \tag{6.25}
$$

where $\tilde{\mathscr{S}}_{\{/i\}}$ are all other fuzzy labels except $\tilde{s}_i$, and $\tilde{\mathscr{N}}_i$ are all neighboring fuzzy labels of $\tilde{s}_i$. The central idea of T2 FMRFs is that we derive clique potentials from T2 MFs for both fuzzy and random uncertainties. The primary MF makes full use of statistical knowledge. The secondary MF is the possibility associated with the primary MF.

The neighborhood system $\tilde{\mathscr{N}}_i$ is defined in Sect. 6.1.1. Without loss of generality, we consider only sing-site and pair-site cliques $\mathscr{C}_1$ and $\mathscr{C}_2$ as mentioned in Sect. 6.1.2. From the posterior energy function (6.5), likelihood energy function (6.3), and prior energy function (6.4), the fuzzy energy function is the joint of the fuzzy likelihood energy function and prior energy function,

$$
\begin{array}{ll}
\textbf{input} & : \mathbf{O}^r = \{\mathbf{o}_1^r, \ldots, \mathbf{o}_I^r\}, 1 \le r \le R, \lambda = \\
& \{\boldsymbol{\mu}_{jm}, \boldsymbol{\Sigma}_{jm}, w_{jm}, \boldsymbol{\mu}_{jj'm}, \boldsymbol{\Sigma}_{jj'm}, w_{jj'm}\}. \\
\textbf{output} & : \hat{\lambda} = \{\hat{\boldsymbol{\mu}}_{jm}, \hat{\boldsymbol{\Sigma}}_{jm}, \hat{w}_{jm}, \hat{\boldsymbol{\mu}}_{jj'm}, \hat{\boldsymbol{\Sigma}}_{jj'm}, \hat{w}_{jj'm}\}.
\end{array}
$$

**1 begin**

**2**     **for** $r \leftarrow 1$ **to** $R$ **do**

**3**     |    $f_{jm}^r(i) \leftarrow \texttt{relaxationlabeling}(\mathbf{O}^r, \lambda);$

**4**     **end**

**5**     $\hat{\boldsymbol{\mu}}_{jm} \leftarrow \dfrac{\sum_{r=1}^{R} \sum_{i=1}^{I_r} f_{jm}^r(i) \mathbf{o}_i^r}{\sum_{r=1}^{R} \sum_{i=1}^{I_r} f_{jm}^r(i)};$

**6**     $\hat{\boldsymbol{\mu}}_{jj'm} \leftarrow \dfrac{\sum_{r=1}^{R} \sum_{i=1}^{I_r} f_{jm}^r(i) \mathbf{o}_{ii'}^r}{\sum_{r=1}^{R} \sum_{i=1}^{I_r} f_{jm}^r(i)};$

**7**     $\hat{\boldsymbol{\mu}}_{jj'm} \leftarrow \dfrac{\sum_{r=1}^{R} \sum_{i=1}^{I_r} f_{jm}^r(i) \mathbf{o}_i^r}{\sum_{r=1}^{R} \sum_{i=1}^{I_r} f_{jm}^r(i)};$

**8**     $\hat{\boldsymbol{\Sigma}}_{jm} \leftarrow \dfrac{\sum_{r=1}^{R} \sum_{i=1}^{I_r} f_{jm}^r(i)(\mathbf{o}_i^r - \hat{\boldsymbol{\mu}}_{jm})(\mathbf{o}_i^r - \hat{\boldsymbol{\mu}}_{jm})'}{\sum_{r=1}^{R} \sum_{i=1}^{I_r} f_{jm}^r(i)};$

**9**     $\hat{\boldsymbol{\Sigma}}_{jj'm} \leftarrow \dfrac{\sum_{r=1}^{R} \sum_{i=1}^{I_r} f_{jm}^r(i)(\mathbf{o}_{ii'}^r - \hat{\boldsymbol{\mu}}_{jj'm})(\mathbf{o}_{ii'}^r - \hat{\boldsymbol{\mu}}_{jj'm})'}{\sum_{r=1}^{R} \sum_{i=1}^{I_r} f_{jm}^r(i)};$

**10**    $\hat{w}_{jm}, \hat{w}_{jj'm} \leftarrow \dfrac{\sum_{r=1}^{R} \sum_{i=1}^{I_r} f_{jm}^r(i)}{\sum_{r=1}^{R} \sum_{i=1}^{I_r} f_j^r(i)};$

**11 end**

**Fig. 6.4** The learning algorithm

$$
\tilde{U}(\mathscr{S}|\mathbf{O}, \lambda) \propto \tilde{U}(\mathbf{O}|\mathscr{S}, \lambda) \sqcup \tilde{U}(\mathscr{S}|\lambda), \tag{6.26}
$$

where

$$
\tilde{U}(\mathbf{O}|\mathscr{S}, \lambda) = \sqcup_{c \in \mathscr{C}_1, \mathscr{C}_2} \tilde{V}_c(\mathbf{O}|\mathscr{S}), \tag{6.27}
$$

$$
\tilde{U}(\mathscr{S}|\lambda) = \sqcup_{c \in \mathscr{C}_1, \mathscr{C}_2} \tilde{V}_c(\mathscr{S}), \tag{6.28}
$$

are a joint of fuzzy clique potentials over all possible cliques $c$.

Similar to Eqs. (6.8) and (6.10), we derive the fuzzy clique potentials from T2 MFs in order to evaluate local neighborhood constraints instead of the probability measure in MRFs. We use the T2 FGMM (4.3) or (4.4) as T2 MFs. By the Hammersley-Clifford theorem as well as the independence assumption (3.69), the fuzzy single-site and pair-site likelihood clique potentials are

$$
\tilde{V}_{\mathscr{C}_1}(\mathbf{o}_i | \tilde{s}_i = j) = -\log \left[ \sqcup_{m=1}^{M_s} \big( w_{jm} \sqcap \tilde{G}(\mathbf{o}_i; \boldsymbol{\mu}_{jm}, \boldsymbol{\Sigma}_{jm}) \big) \right], \tag{6.29}
$$

$$
\tilde{V}_{\mathscr{C}_2}(\mathbf{o}_i, \mathbf{o}_{i'} | \tilde{s}_i = j, \tilde{s}_{i'} = j', i \in \mathscr{N}_i) = -\log \left[ \sqcup_{m=1}^{M_s} \big( w_{jj'm} \sqcap \tilde{G}(\mathbf{o}_{ii'}; \boldsymbol{\mu}_{jj'm}, \boldsymbol{\Sigma}_{jj'm}) \big) \right], \tag{6.30}
$$

where $\mathbf{o}_i$ and $\mathbf{o}_{ii'}$ are unary and binary features respectively. The fuzzy single-site prior clique potential is a penalty for null labels without assigning to any observations denoted by $j = 0$,

$$\tilde{V}_{\mathscr{C}_1}(j) = \begin{cases} -\tilde{v}_1, & \text{if } j \neq 0 \\ \tilde{v}_1, & \text{if } j = 0, \end{cases} \tag{6.31}$$

and pair-site prior clique potential

$$\tilde{V}_{\mathscr{C}_2}(j, j') = \begin{cases} -\tilde{v}_2, & j \text{ is connected with } j', \\ \tilde{v}_2, & j \text{ is disconnected with } j'. \end{cases} \tag{6.32}$$

where $\tilde{v}_1$ and $\tilde{v}_2$ are T1 fuzzy numbers with positive support $v_1$ and $v_2$, which reflect the fuzziness of the local labeling constraints. The connection of labels $j$ and $j'$ is defined in Sect. 6.1.2.

For the computational efficiency, we use the IT2 MFs leading to the IT2 FMRFs. Hence, we can replace the "$\sqcup$" with bounded "$\sum$", and replace the "$\sqcap$" with $t$-norm "$\star$" from Eqs. (6.26)–(6.30). In this case, the fuzzy likelihood clique potentials are interval sets. Furthermore, the fuzzy prior clique potentials, $\tilde{v}_1$ and $\tilde{v}_2$, become positive constant interval sets, $\tilde{v}_1 = [\underline{v}_1, \overline{v}_1]$ and $\tilde{v}_2 = [\underline{v}_2, \overline{v}_2]$.

### 6.2.1  The Type-2 Fuzzy Relaxation Labeling

To find the best fuzzy labeling configuration $\mathscr{S}^*$ and the minimum fuzzy energy $U_{\tilde{\lambda}}(\mathscr{S}^* | \mathbf{O})$, we use the IT2 fuzzy relaxation labeling algorithm as shown in Fig. 6.5. We define $\tilde{f}_j(i)$ as the IT2 membership grade of $\tilde{s}_i = j$ given observations $\mathbf{O}$, i.e., $\tilde{f}_j(i) = h_{\tilde{\lambda}}(\tilde{s}_i = j | \mathbf{O})$. Similar to the gain functions (6.18) and (6.19), we convert the minimization of fuzzy energy to maximize the following fuzzy gain functions,

$$g_{\underline{\tilde{\lambda}}}(\mathscr{S} | \mathbf{O}) = \sum_{i=1}^{I} \sum_{j=1}^{J} \left[ \tilde{K}_j(i) \star \tilde{f}_j(i) + \sum_{i' \in \mathscr{N}_i} \sum_{j'=1}^{J} \tilde{K}_{j,j'}(i, i') \star \tilde{f}_j(i) \star \tilde{f}_{j'}(i') \right], \tag{6.33}$$

and

$$g_{\overline{\tilde{\lambda}}}(\mathscr{S} | \mathbf{O}) = \sum_{i=1}^{I} \sum_{j=1}^{J} \left[ \tilde{K}_j(i) \star \tilde{f}_j(i) + \sum_{i' \in \mathscr{N}_i} \max_{j'=1}^{J} \tilde{K}_{j,j'}(i, i') \star \tilde{f}_j(i) \star \tilde{f}_{j'}(i') \right]. \tag{6.34}$$

The compatibility functions are defined by fuzzy clique potentials,

$$
\begin{array}{ll}
\textbf{input} & : \mathbf{O} = \{\mathbf{o}_1, \ldots, \mathbf{o}_I\}, \lambda = \{a_{jj'}, \tilde{V}_{C_1}, \tilde{V}_{C_2}\}. \\
\textbf{output} & : \tilde{S}^*, \tilde{U}(\tilde{S}^* | \mathbf{O}, \lambda). \\
\textbf{initialize} & : \tilde{f}_j^1(i) \leftarrow \frac{1}{J}, \tilde{K}_j(i) \leftarrow CONST_1 - \tilde{V}_{C_1}(\mathbf{o}_i | \tilde{s}_i = \\
& \quad j) - \tilde{V}_{C_1}(j), \tilde{K}_{j,j'}(i, i') \leftarrow \\
& \quad CONST_2 - \tilde{V}_{C_2}(\mathbf{o}_i, \mathbf{o}_{i'} | s_i = j, s_{i'} = j', i' \in \\
& \quad \mathcal{N}_i) - \tilde{V}_{C_2}(j, j'), 1 \le j, j' \le J, 1 \le i \le I, i' \in \mathcal{N}_i.
\end{array}
$$

```
1  begin
2  |   for t ← 1 to T_max do
3  |   |   for i ← 1 to I do
4  |   |   |   for j ← 1 to J do
5  |   |   |   |   q̃_i^t(j) ←
       K̃_j(i) + Σ_{i'∈𝒩_i} Σ_{j'=1}^{J} K̃_{j,j'}(i,i') ⋆ f̃_{j'}(i')  or
       K̃_j(i) + Σ_{i'∈𝒩_i} max_{j'=1}^{J} K̃_{j,j'}(i,i') ⋆ f̃_{j'}(i');
6  |   |   |   |   f̃_j^{t+1}(i) ← (f̃_j^t(i) ⋆ q̃_j^t(i)) / (Σ_{j=1}^{J} f̃_j^t(i) ⋆ q̃_j^t(i));
7  |   |   |   end
8  |   |   end
9  |   end
10 |   for i ← 1 to I do
11 |   |   s̃_i^* = arg max_{j=1}^{J} f̃_j^T(i);
12 |   end
13 |   Ũ_λ̄(S^*|O) ← CONST − g_λ̄(S̃^*|O);
14 end
```

**Fig. 6.5**  The interval type-2 fuzzy relaxation labeling algorithm

$$
\tilde{K}_j(i) = \mathrm{CONST}_1 - \tilde{V}_{\mathscr{C}_1}(\mathbf{o}_i | \tilde{s}_i = j) - \tilde{V}_{\mathscr{C}_1}(j),
$$
$$
\tilde{K}_{j,j'}(i, i') = \mathrm{CONST}_2 - \tilde{V}_{\mathscr{C}_2}(\mathbf{o}_i, \mathbf{o}_{i'} | s_i = j, s_{i'} = j', i' \in \mathcal{N}_i) - \tilde{V}_{\mathscr{C}_2}(j, j').
$$
$$(6.35)$$

The constant $\mathrm{CONST}_1$ and $\mathrm{CONST}_2$ satisfy that all the compatibility functions are non-negative.

Then we update $\tilde{f}_i^t(j)$ by the gradient of the fuzzy gain functions (6.33) and (6.34),

$$
\tilde{q}_j(i) = \tilde{K}_j(i) + \sum_{i' \in \mathcal{N}_i} \sum_{j'=1}^{J} \tilde{K}_{j,j'}(i, i') \star \tilde{f}_{j'}(i'),
$$
$$
\tilde{q}_j(i) = \tilde{K}_j(i) + \sum_{i' \in \mathcal{N}_i} \max_{j'=1}^{J} \tilde{K}_{j,j'}(i, i') \star \tilde{f}_{j'}(i'),   \qquad (6.36)
$$

until $t$ reaches a fixed number $T_{\max}$ as shown in Fig. 6.5. Because the IT2 FMRF compatibility function contains both fuzzy likelihood and prior information, the solution of relaxation labeling dose not depend on the initial labeling that much. Finally, we use the winner-take-all strategy to retrieve the best fuzzy labeling configuration,

$$\tilde{s}_i^* = \arg\max_{j=1}^{J} \tilde{f}_j(i). \tag{6.37}$$

We estimate parameters of IT2 FMRFs by initialization and Baum-Welch algorithms as shown in Figs. 6.3 and 6.4. Note that we use the IT2 relaxation algorithm to align the observations with labels, and the updating weight in Baum-Welch is the centroid of $\tilde{f}_j(i)$. The uncertainty factor $k$ or $k_1$, $k_2$, and the range of the interval sets $\tilde{v}_1$ and $\tilde{v}_2$ are determined based on prior knowledge.

### 6.2.2 Computational Complexity

The classical RL algorithm has polynomial complexity in Sect. 6.1.3. If we choose bounded sum $t$-conorm in the join operation and the product $t$-norm in the *meet* operation, the IT2 fuzzy RL procedure for IT2 FMRFs can be viewed as computing the best labeling of two boundary MRFs: the "lower" MRF and the "upper" MRF. Therefore, the computational complexity of IT2 FMRFs is twice that of the classical MRFs.

## 6.3 Stroke Segmentation of Chinese Character

In this section, we present MRFs to segment strokes of Chinese characters. The distortions caused by the thinning process make the thinning-based stroke segmentation difficult to extract continuous strokes and handle the ambiguous intersection regions. The MRF reflects the local statistical dependencies at neighboring sites of the stroke skeleton, where the likelihood clique potential describes the statistical variations of directional observations at each site, and the smoothness prior clique potential describes the interactions among observations at neighboring sites. Based on the cyclic directional observations by Gabor filters, we formulate the stroke segmentation as an optimal labeling problem by the *maximum a posteriori* (MAP) criterion. The results of stroke segmentation on the ETL-9B character database are encouraging.

A Chinese character is constituted by a sequence of straight-line stroke segments with different positions, directions, and lengths, which play important roles in character recognition [1]. According to the structure of Chinese characters, most stroke segmentation methods decompose characters into four-directional stroke segments: horizontal (0°), right-diagonal (45°), vertical (90°), and left-diagonal (135°) [2, 3]. Traditional stroke segmentation methods by thinning process [4, 5] is much easier to implement with lower computational complexity than those without thinning process [2, 3, 6, 7]. However, thinning process distorts the character shape especially at the intersection regions as shown in Fig. 6.6. Such distortions cause two types of ambiguous strokes: one is the sharing part by more than two strokes at the intersection region; the other is the transition part at the high curvature region

**Fig. 6.6** The thinned character image is on the *right*. The ambiguous parts due to distortions are marked by the *red stars*

of the continuous stroke. These ambiguous parts will cause broken strokes, which increases the complexity of the character recognition system. This section formulates the stroke segmentation as an optimal labeling problem based on MRFs, in which each site on the thinned character is labeled with one of the four directional labels 1–4, or the ambiguous label 5.

Within the MRF framework for the low-level computer vision problem, the local statistical interactions among adjacent sites in a pattern or image are reflected by two fundamental concepts: *neighborhood system $\mathcal{N}$* and *clique potentials $V_c$*. To encourage or penalize different local interactions among neighboring sites, we assign different costs $V_c$ to cliques. By the equivalence between MRFs and GRFs (Gibbs random fields) [8], computing MAP of the MRF guarantees the best labeling configuration [9].

In light of the MAP-MRF framework in Sect. 2.4, let us now reexamine the stroke segmentation. Each site $i$ on the character is associated with a Gabor filters-based observation $\mathbf{o}_i$ that can be modeled by the conditional probability $p(\mathbf{o}_i|s_i = j)$, $1 \leq j \leq 5$. To obtain continuous stroke segments, we impose the smoothness constraints to the spatial neighboring sites, because if one observation $\mathbf{o}_i$ is labeled with $j$, the neighboring observation $\mathbf{o}_{i'}$ is more likely labeled with $j$. This prior can be modeled by the joint probability $P(\mathcal{S})$. Finally, we use the relaxation labeling (RL) algorithm to search the best labeling configuration $\mathcal{S}^*$ for the directional observations $\mathbf{O}$ according to the MAP criterion. Therefore, this MAP-MRF framework not only improves the accuracy of segmentation by modeling various relationships among directional observations at neighboring sites statistically, but also offers rational principles for segmentation rather than *ad hoc* postprocessing heuristics in [2–7].

### 6.3.1 Gabor Filters-Based Cyclic Observations

To extract the stroke is very difficult due to complicated character shapes, varied stroke widths, and the ambiguous stroke intersections. We use 2D Gabor filters due to following attractive qualities: (1) Gabor filters can extract various kinds of visual features, including lines, shapes, and textures, which is especially suited to regular-style characters composed of straight-line segments; (2) The responses of Gabor filters are robust under small translation, rotation, and scaling of the objects; (3) The Gaussian nature of the filters makes them tolerant to noise.

The 2D Gabor filter in space domain is a 2D Gaussian-shaped envelop modulated a complex sinusoidal carrier [10],

$$g(x, y) = K \exp\left(-\pi \left(\frac{(x - x_0)_r^2}{\sigma_x^2} + \frac{(y - y_0)_r^2}{\sigma_y^2}\right)\right) \exp(j(2\pi F(u_0 x + v_0 y) + P)),$$

(6.38)

where

$$(x - x_0)_r = (x - x_0) \cos \omega + (y - y_0) \sin \omega,$$  (6.39)
$$(y - y_0)_r = -(x - x_0) \sin \omega + (y - y_0) \cos \omega.$$  (6.40)

If the sinusoidal carrier is expressed in polar coordinates as magnitude $f$ and orientation $\theta$,

$$f = \sqrt{u_0^2 + v_0^2},$$  (6.41)

$$\theta = \tan^{-1}\left(\frac{v_0}{u_0}\right),$$  (6.42)

i.e.

$$u_0 = f \cos \theta,$$  (6.43)
$$v_0 = f \sin \theta,$$  (6.44)

then the 2D Gabor filter can be rewritten as

$$g(x, y) = K \exp\left(-\pi \left(\frac{(x - x_0)_r^2}{\sigma_x^2} + \frac{(y - y_0)_r^2}{\sigma_y^2}\right)\right) \exp(j(2\pi f(x \cos \theta + y \sin \theta) + P)).$$

(6.45)

The 2D Gabor filter is defined by the following parameters.
$K$        Scales the magnitude of the Gaussian envelop.
$(\sigma_x, \sigma_y)$ Standard deviation of the Gaussian envelop along $x$- and $y$-axis.
$\omega$        Rotation angle of the Gaussian envelop.
$(x_0, y_0)$ Location of the peak of the Gaussian envelop.
$(u_0, v_0)$ Spatial frequency of the sinusoidal carrier in Cartesian coordinates.
$(f, \theta)$   The frequency and the orientation of Gabor filter.
$P$        Phase of the sinusoidal carrier.
    To extract directional observations, we simplify the 2D Gabor filter according to the following constraints [2, 10–13]: (1) The phase $P = 0$, the magnitude $K = 1$ and the location of the filter center ($x_0 = 0$, $y_0 = 0$); (2) The standard deviation $\sigma_x = \sigma_y = \sigma$, where $f$ is the frequency of the Gabor filter; (3) The rotation of Gaussian envelop equals the orientation of the sinusoidal carrier, i.e., $\omega = \theta$. The

Eq. (6.38) can be rewritten as

$$g(x, y) = \exp\left(-\pi \left(\frac{x^2 + y^2}{\sigma^2}\right)\right) \exp\left(j2\pi f x'\right), \qquad (6.46)$$

where

$$x' = x \cos\theta + y \sin\theta. \qquad (6.47)$$

Let $i(x, y)$ denote an input binary character image, and let $I(x, y)$ denote the output of the 2D Gabor filter. Then, the output is the magnitude of the convolution between input image and the 2D Gabor filter,

$$I(x, y) = |i(x, y) * g(x, y)|, \qquad (6.48)$$

where the Gabor filter template is a square with $x \in [-\sigma_x, \sigma_x]$ and $y \in [-\sigma_y, \sigma_y]$.

For the selection of filter frequency, Su proposed the relationship between filter frequency and the character complexity [2],

$$f = \frac{\alpha SW}{DH} + \beta, \qquad (6.49)$$

where $SW$ is the stroke width, $H$ denotes the image height, $D$ represents the character complexity as measured by the ratio of the number of stroke pixels to the image size. Herein, the additional parameters $\alpha$ and $\beta$ are 0.9875 and 0.12. The stroke width $SW$ is approximated by

$$SW = \frac{\text{the area of character}}{\text{the edge of character}}, \qquad (6.50)$$

where the area and the edge are the number of pixels of the character and the character edge, respectively. In thinned character images, we have $SW \approx 1$. The standard deviation of the Gaussian envelop $\sigma = \sqrt{2}/f$ [2, 14]. The Gabor filter is robust to the selection of $f$ and $\sigma$, i.e., the outputs are stable with small variations of $f$ and $\sigma$.

In the literature [15], the directional observation $\mathbf{o}^D$ of the stroke is defined in an interval $[-90, 270°]$, which is not cyclic because different values may represent almost the same direction, e.g., $-89$ and $269°$ are almost the same direction of the stroke. To extract cyclic $\mathbf{o}^D$, we use 2D Gabor filters [2, 16, 17] as follows.

1. We use eight Gabor filters with orientations 0, 22.5, 45, 67.5, 90, 112.5, 135, and 157.5° to convolve with the character skeleton image, which results in eight uncorrelated gray images.
2. At each pixel on the character skeleton from eight gray images, we get eight independent values that make up $\mathbf{o}^D$. We normalize it further by setting its maximum and minimum values to one and zero, such as $\mathbf{o}^D = [1.00, 0.65, 0.05, 0.02, 0, 0.04, 0.09, 0.58]'$.

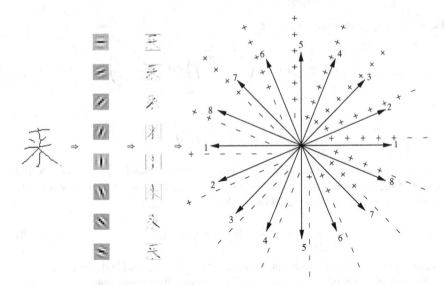

**Fig. 6.7** Extract cyclic directional observations $\mathbf{o}^D$ by 2D Gabor filters. To reflect the relative position to the origin, we assign plus and minus signs to elements of $\mathbf{o}^D$

3. The $\mathbf{o}^D$ of a substroke is the average of $\mathbf{o}^D$ of its component pixels. If a substroke has $\mathbf{o}^D = [0.81, 0.83, 0.28, 0.06, 0.12, 0.10, 0.36, 0.59]'$, it must be a horizontal line because its responses of Gabor filters are larger in 0 and 22.5° orientations.
4. To reflect the relative position to the origin, we assign plus and minus signs to elements of $\mathbf{o}^D$ as shown in Fig. 6.7. For example, the substroke lies between directions 1 and 2 and upper-right to the origin with $\mathbf{o}^D = [0.81, 0.83, 0.28, 0.06, 0.12, -0.10, -0.36, -0.59]'$.

Figure 6.7 shows the process of directional observation extraction by Gabor filters. From any start point, the $\mathbf{o}^D$ changes continuously in both clockwise and counterclockwise directions, which demonstrates that the $\mathbf{o}^D$ is the cyclic directional observation.

Here, we take only the first two steps to extract directional observation at each pixel $i$ (site). We will use the last two steps to extract the substroke directional observation for handwritten Chinese character recognition.

### 6.3.2 Stroke Segmentation Using MRFs

The idea is that we design the neighborhood system and clique potentials of MRFs for the optimal labeling of the character skeleton in terms of four directional labels 1–4, and the ambiguous label 5.

**Fig. 6.8** The neighborhood system

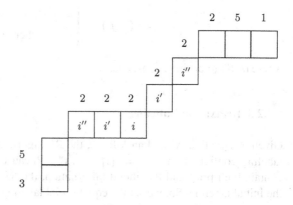

### 6.3.2.1 Neighborhood Systems

We define the neighborhood system based on the connection of sites as shown in Fig. 6.8. The sites $i'$ and $i''$ are the first-order and second-order neighbors of the site $i$ respectively. The $N$th-order neighborhood system can be defined in the same way.

### 6.3.2.2 Clique Potentials

We represent $p(\mathbf{o}_i|s_i = j)$ in Eq. (2.14) by a multivariate Gaussian (5.4) with mean vector $\boldsymbol{\mu}_j$ and covariance matrix $\boldsymbol{\Sigma}_j$, and derive the single-site likelihood clique potential (6.8). The estimates of $\boldsymbol{\mu}_j$ and $\boldsymbol{\Sigma}_j$ are the following simple averages in terms of training observations $\mathbf{o}^r$, $1 \le r \le R$,

$$\hat{\boldsymbol{\mu}}_j = \frac{\sum_{r=1}^{R} \mathbf{o}^r}{R}, \tag{6.51}$$

$$\hat{\boldsymbol{\Sigma}}_j = \frac{\sum_{r=1}^{R} (\mathbf{o}^r - \boldsymbol{\mu}_j)(\mathbf{o}^r - \boldsymbol{\mu}_j)'}{R}. \tag{6.52}$$

Similarly, we derive the pair-site likelihood clique potential (6.10), where the binary observation $\mathbf{o}_{ii'} = \mathbf{o}_{i'} - \mathbf{o}_i$. The parameters are estimated by Eqs. (6.51) and (6.52) using the binary training observations.

The single-site prior clique potential (6.12) does not work in the stroke segmentation, so we set them zero,

$$V_{\mathscr{C}_1}(j) = 0. \tag{6.53}$$

We design the pair-site smoothness prior clique potential (6.14) as follows,

$$V_{\mathcal{C}_2}(j, j') = \begin{cases} -\alpha, & \text{if } j = j', \\ \beta, & \text{otherwise,} \end{cases} \qquad (6.54)$$

where $\alpha, \beta$ are positive constants.

### 6.3.2.3 Relaxation Labeling

Given observations $\mathbf{O}$ and an MRF $\lambda$, the RL algorithm (Fig. 6.2) can find the best labeling configuration, $\mathscr{S}^* = \{s_1^*, \ldots, s_j^*\}$. Because the compatibility functions contain both prior and likelihood information, the RL does not heavily depend on the initial labeling. So, we set the equal initial labeling $f_j^1(i) = 0.2, 1 \leq j \leq 5$.

## 6.3.3 Stroke Extraction of Handprinted Chinese Characters

Experiments were carried out on the handprinted Chinese character database ETL-9B, which includes 2,965 classes of Chinese characters with binary files of $63 \times 64$ pixels resolution. We used 50 thinned characters as training data, and manually labeled the ambiguous parts at intersection and transition regions. We also selected 500 characters for test, among which 463 characters were correctly segmented, and the rest 37 had broken strokes. The correct rate is 92.6 % and the complexity is about 8 strokes per character on the average. Some results are shown in Fig. 6.9, where the ambiguous parts are marked by red stars, confirming that MRFs can effectively detect ambiguous parts. The characters on the sixth and seventh rows have broken strokes in Fig. 6.18. The reason for broken strokes is because MRFs' local rather than global probability measure of dependencies among neighboring sites on the character skeleton. Although some postprocessing rules can be added to connect the broken strokes, they may also connect two different strokes together.

For simplicity, we use only the first-order neighborhood system. We believe a high-order (larger) neighborhood system with more complicated cliques will make the stroke segmentation better. Indeed, the stroke segmentation results are highly dependent on the directional observations $\mathbf{o}^D$. When the stroke direction does not fall into four directions, we have to use another method to extract reliable strokes, such as model-based method in handwritten Chinese character recognition.

## 6.3.4 Stroke Extraction of Cursive Chinese Characters

We focus on the stroke segmentation of cursive Chinese characters because it is still a challenging problem. We performed extensive experiments on KAIST Hanja1/Hanja2 handwritten Chinese character databases [18]. Hanja1 has 783 categories with 200 samples for each category. Hanja2 has 1309 samples of naturally

**Fig. 6.9** Stroke segmentation results

cursive Chinese characters, which are most difficult in segmentation because of touched strokes. The Hanja1 image quality is good, but Hanja2 is bad. Direct usage of cursive characters as training data is infeasible, because cursive samples have large variations in shape. So, from Hanja1 database, we used 50 thinned characters as training data, and manually assigned ambiguous labels at junction and transition regions. For test purpose on Hanja2, we segmented all 1309 cursive characters with the complexity about nine strokes per character on the average. We compared cascade MRFs with a recent method based on Gabor filters referred to as GF method [2]. In this stroke extraction method, a set of Gabor filters is used to break down an image of

**Fig. 6.10** Stroke segmentation results of cursive Chinese characters on KAIST Hanja2 database. The *horizontal*, *left-diagonal*, *vertical*, *right-diagonal*, and ambiguous labels are denoted by ∗, +, ×, ▽, and ∘, respectively. The first line is produced by the MRF and the second line is produced by the GF method [2]. Some fragmented substrokes are marked by *circles*

a character into four directional features, and then an iterative thresholding technique is used to recover stroke shape by minimizing the reconstruction error. A refinement process is used to remove redundant stroke pieces based on measuring the degree of stroke overlap. This method has been confirmed to be effective on well-written Chinese characters, but has not been examined on cursive Chinese characters.

Figure 6.10 illustrates results of stroke segmentation of cursive Chinese characters on Hanja2, where the first line is produced by the MRF and the second line is produced by the GF method [2]. Two important observations are as follows. First, the MRF can effectively detect ambiguous parts labeled by the symbol o. These ambiguous parts play important roles in producing continuous substrokes because two substrokes concatenated with ambiguous parts can belong to the same substroke if they are associated with the same directional label. Second, the GF method produces more fragmented substrokes than those produced by the MRF because cursive Chinese characters often have less straight-line strokes than well-written Chinese characters. For the GF method, we marked those fragmented substrokes by circles in Fig. 6.10. These fragmented noisy substrokes will increase the complexity in the merging process to produce perceptually meaningful strokes. Particularly, fragmented substrokes cause broken strokes that deteriorate the Chinese character structure significantly. In contrast, by introducing the smoothness-based prior, the MRF performs well in retaining smoothness of substrokes as much as possible. Furthermore, we introduce a statistical learning of directional features of substrokes so as to handle local shape variations of cursive Chinese characters. Another advantage is that the MRF-based stroke segmentation does not require an external corner detection [15, 19], or line approximation [4, 18], to break the substroke at the high curvature places.

Figure 6.11 illustrates the stroke extraction results of cursive characters on Hanja2 database. The first column is the MRF-based character models composed of many stroke labels. The second column shows stroke extraction results by the MRF. Only the best candidate strokes with the lowest clique potentials are illustrated. The numbers denote the correspondence between stroke labels and extracted strokes. The third column shows stroke extraction results by the GF method. For the GF method, we marked broken strokes by "B" and touching strokes by "T," where broken strokes mean that substrokes are not correctly merged and touching strokes mean that substrokes are incorrectly merged. For the MRF-based character model, we see that the stroke labels can accurately find the perceptually meaningful candidate strokes in the second column. Particularly, stroke labels can guide the merging of proper substrokes and break some touching strokes in cursive characters, which cannot be achieved by the GF method in the third column. Indeed, the GF method almost fail to produce reliable strokes because it lacks the global information such as position and length of the original strokes. This condition is more serious in cursive Chinese characters due to unreliable local information. To summarize, it is advantageous to combine both BU and TD vision processing streams together to extract perceptually meaningful strokes from cursive Chinese characters.

## 6.4 Handwritten Chinese Character Recognition

In this section, we propose a statistical-structural strategy for Chinese character modeling and recognition based on MRFs in Sect. 2.4. The relationships between strokes of a Chinese character reflect its structure, which can be statistically represented by the neighborhood system and clique potentials within the MRF framework. To handle

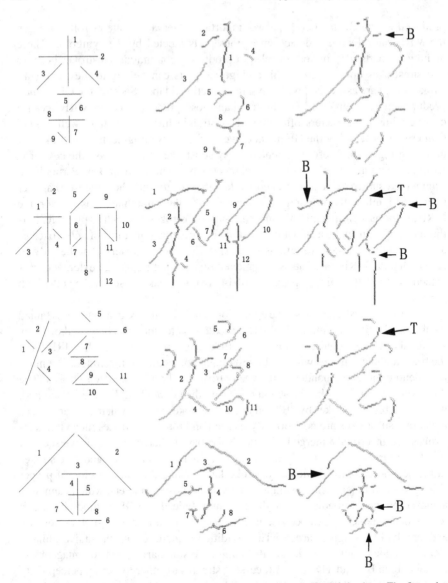

**Fig. 6.11** Stroke extraction results of cursive Chinese characters on Hanja2 database. The first column is the MRF-based character models composed of many stroke labels. The second column shows stroke extraction results by the MRF. Only the best candidate strokes with the lowest clique potentials are illustrated. The numbers denote the correspondence between stroke labels and extracted strokes. The third column shows stroke extraction results by the GF method. The broken strokes and touching strokes are marked by "B" and "T," respectively

**Fig. 6.12** The relative position and length between strokes marked by *dotted ellipse* can differentiate these similar characters

large structural variations of the character, we design the neighborhood system as well as the prior clique potential based on prior knowledge of character structures, and derive the likelihood clique potential from Gaussian mixture models. Therefore, in the MRF-based character model, the likelihood clique potential describes the relationships of strokes statistically, and the prior clique potential encodes the prior structural information of characters in the neighborhood system. We apply MRFs to the handwritten Chinese character recognition (HCCR) system, where we use MRFs to extract reliable strokes from images of characters. The experiments on the ETL-9B and KAIST character databases demonstrate that MRFs can represent both statistical and structural information of characters, and work well for HCCR.

The structure of Chinese characters is hierarchical: Many straight-line strokes constitute independent radicals, which in turn constitute characters [20]. According to the theory of structural representation, character shapes can be represented by fragmental features (such as the position of a stroke) and configurational features for relationships among the fragmental features. The human visual system uses mostly configurational features rather than fragmental features to recognize characters during reading [16]. Therefore, character structures play important roles in recognition, especially for similar characters as shown in Fig. 6.12. Because Chinese characters have hierarchical parts with complicated shape information, modeling character structure is one of the most challenging topics in pattern recognition [18].

The statistical and the structural approaches are two important strategies for modeling characters [15, 16, 18, 20–22]. The statistical approaches can represent character structures indirectly by feature vectors for the holistic shape information. With such a representation, we can use standard statistical methodologies to recognize characters (e.g., city block distance and Mahalanobis distance [23], k-nearest-neighborhood classifier [24], k-means clustering and Gaussian distribution selector [25], and contextual vector quantization [26]). On the other hand, the structural approaches can represent the fine detail of character structures by a character model composed of a set of stroke models corresponding to real strokes. The structural matching is then performed between input strokes and stroke models for character recognition. Generally, the structural approaches extract feature points and line segments from the character, and represent their relationships by a relational graph, in which the nodes represent feature points or line segments, and the edge between two nodes for their relationships (e.g., constraint graph model [21], attributed relational graph [15], and hierarchical random graph [27]). The statistical approaches have systematical learning schemes from training samples, but reflect the character structure indirectly [16, 18, 20]. In contrast, the structural approaches can represent the character structure sufficiently, but depend heavily on the developers knowledge and reliable stroke extraction, which leads to neither rigorous matching algorithm nor automatic leaning schemes from training samples [15, 18, 27]. Therefore, a theoretically well-founded approach is needed to represent both statistical and structural information of characters within a unified framework.

MRFs can represent 2D structural patterns statistically. Their great success achieved in pattern recognition, image processing, and computer vision in the passing decades have been largely due to their ability to reflect local statistical dependencies existing universally in patterns, images, and video frames [9, 28, 29]. Within the MRF framework, statistical interactions at adjacent sites in a pattern or image are reflected by two fundamental concepts: *neighborhood system* $\mathcal{N}$ and *clique potentials* $V_c$. We can define two sites $i$ and $i'$ as neighbors to each other in $\mathcal{N}$ if these two sites have relationships. The clique is a subset of sites that are all pairwise neighbors. We assign costs $V_c$ to different cliques to encourage or penalize different local interactions among neighboring sites. By using the equivalence between MRFs and GRFs due to the Hammersley-Clifford theorem [8], we can compute the best global configuration based on the *maximum a posteriori* (MAP) criterion [9].

Let us now reexamine the statistical character structure modeling. The stroke is random in terms of its direction, position, and length, which can be described by PDFs (probability density functions) at each site. Different stroke relationships identify different character structures, which can be reflected by interactions among PDFs at neighboring sites in the neighborhood system $\mathcal{N}$. Within the MAP-MRF framework, we can encourage or penalize different stroke relationships by assigning different clique potentials to spatial neighboring sites. These clique potentials can be derived from PDFs, and their parameters can be estimated from training samples automatically. In the MRF-based statistical character structure modeling, we focus on the following three central issues: (1) Define the neighborhood system that accounts for the most important stroke relationships; (2) Design clique potentials that evaluate various local statistical dependencies among strokes; (3) Extract reliable strokes from images of characters. Not only dose the MAP-MRF framework improve the accuracy of structural matching between input strokes and character models, but also offer rational principles to learn from training samples automatically rather than *ad hoc* heuristics.

Unlike the rigorous MRF definition, the hidden MRF [30] and contextual stochastic modeling [31] mainly focus on the causal dependencies in the neighborhood system $\mathcal{N}$, and use the dynamic programming to alleviate practical computational cost. However, noncausal stroke relationships are more reasonable because the temporal stroke-order information is unknown in off-line character recognition. Statistical character structure modeling (SCSM) [18] describes the stroke relationships by the conditional probability of the neighbors selected by minimizing *Kullback-Leibler measure*, and uses heuristic search to find the best correspondence between input strokes and stroke models. SCSM defines the neighborhood system by *Kullback-Leibler measure* that may be viewed as a special case of the

MRF if the joint probability of strokes is represented by GRFs. Similarly, stochastic modeling of stroke relationships (SMSR) [22] can be considered within the MRF framework if it directly uses the GRF to model stochastic relationships. Compared with above statistical character structure modeling strategies, we believe that the theoretically well-founded MAP-MRF may shed more light on building a salient framework for statistical character structure modeling and recognition.

## 6.4.1 MRFs for Character Structure Modeling

Character recognition can be viewed as a labeling problem, and the solution to this problem is a set of labels assigned to the input strokes. An MRF for each class of characters is a character model composed of many labels referred to as the stroke models. Each label is associated with a random variable for a stroke segment. Different configurations of labels identify different stroke relationships. Every two labels being neighbors with a possibility defines a global neighborhood system to describe complicated character structures. Clique potentials are derived from Gaussian mixture densities, whose parameters can be estimated from training data automatically by generalized EM algorithm. The labels assignment depends on the neighborhood system and clique potentials, and the classical relaxation labeling can assign labels to the stroke segments efficiently. This MRF-MAP framework can also incorporate our prior knowledge into matching process in terms of the prior clique potentials (6.4).

### 6.4.1.1 Stroke Features

The direction, position, and length are the complete spatial information of strokes. At each site $i$, we represent this information by the observation vector, $o_i = (o_i^D, o_i^P, o_i^L)'$, which is the unary feature of an individual stroke. The relationship between two neighboring strokes can be represented by the binary feature, $o_{ii'} = (o_{ii'}^D, o_{ii'}^P, o_{ii'}^L)'$, where $o_{ii'}^D = o_{i'}^D - o_i^D$, $o_{ii'}^P = o_{i'}^P - o_i^P$, and $o_{ii'}^L = o_{i'}^L - o_i^L$. The long strokes and short strokes of Chinese characters play different roles in consti-

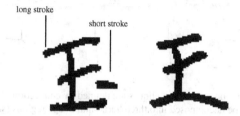

**Fig. 6.13** Three long *horizontal strokes* and one *vertical stroke* constitute the main body of the character "玉", but *short strokes* marked by *dotted ellipse* are crucial to differentiate it from the character "王"

tuting the character structure. The long strokes usually constitute the basic structure of the character, but some short strokes are crucial to differentiate similar characters as shown in Fig. 6.13. Therefore, after relaxation labeling, best labels should be assigned to long strokes before short strokes, and null labels should be imposed to a penalty, especially those for crucial short strokes. We can incorporate this prior knowledge of strokes in the MRF-based character models. To reflect the length information of strokes, we define $\gamma_i$ as

$$\gamma_i = \frac{\mathbf{o}_i^L}{\sum_{i=1}^{I} \mathbf{o}_i^L},\tag{6.55}$$

which shows that the bigger $\gamma_i$ the longer stroke.

Within the MAP-MRF framework, we model these features statistical structurally by the neighborhood system and clique potentials. Meanwhile, we must extract reliable strokes that best represent character structures. In the following sections, we will focus on three central issues: (1) define the neighborhood system, (2) design clique potentials, and (3) extract reliable strokes.

### 6.4.1.2 The Neighborhood System

For high-level vision problems, the neighborhood system is usually defined on irregular sites such as image regions. Two image regions $i$ and $i'$ are neighbors if they are connected as shown in Fig. 6.14. Chinese characters can be decomposed into image regions such as strokes. The strokes that share the same intersection region often have stable structures, such as stable relative directions, positions, and lengths. Thus, the neighborhood system based on connection represents the most important stroke relationships in Chinese characters. Meanwhile, the neighborhood system based on connection also satisfies that we consider only single-site and pair-site cliques in Sect. 2.2, because few Chinese characters have more than two strokes connecting at the same intersection region after the thinning process. Therefore, we can ignore high-order cliques in modeling stroke relationships.

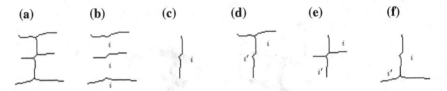

**(a)**      **(b)**      **(c)**      **(d)**      **(e)**      **(f)**

**Fig. 6.14** The neighborhood system and cliques are defined by connection of image regions. The character "王" in **a** can be decomposed into three horizontal image regions (strokes) in **b** and one *vertical* image region (stroke) in **c**. The single-site cliques $\mathscr{C}_1 = \{i\}$ are shown in **b** and **c**, and the pair-site cliques $\mathscr{C}_2 = \{(i, i')\}$ are shown in **d**–**f** respectively

Figure 6.14 shows an example of the neighborhood system and cliques defined by connection. The character "王" has been decomposed into three horizontal strokes and one vertical stroke. We model the most important relationships between these four strokes based on connection, in which only strokes $i$ and $i'$ are neighbors. We can automatically obtain such neighboring information between strokes by the MRF-based stroke extraction in the following sections.

### 6.4.1.3 Clique Potentials

We use the likelihood potential to encode both statistical and structural information of strokes from training samples, and the prior potential to encode the prior structural information based on prior knowledge.

The single-site (6.8) and pair-site (6.10) likelihood potentials are presented in Sect. 5.1.2. To incorporate the length information of strokes, we use the single-site likelihood potential, $\gamma_i V_{\mathscr{C}_1}(\mathbf{o}_i|s_i = j, \lambda)$, which makes the labeling strength of longer strokes reach maximum faster than that of shorter strokes in RL algorithm in Fig. 6.2. For the same reason, we incorporate the length information by using the pair-site likelihood potential, $\gamma_i \gamma_{i'} V_{\mathscr{C}_2}(\mathbf{o}_i, \mathbf{o}_{i'}|s_i = j, s_{i'} = j', \lambda)$.

The prior clique potential reflects the local structure in the labeling space by encouraging the connected labels or penalizing the disconnected labels at neighboring sites. We say that the label $j$ is connected with $j'$ denoted by $a_{jj'} = P(s_{i'} = j'|s_i = j) > 0$, and disconnected if $a_{jj'} = 0$. Note that the label $j$ is always disconnected with itself, i.e., $a_{jj} = 0$. We denote the connection of labels $j$ and $j'$ by an edge. As shown in Fig. 6.15d, label 4 is connected with labels 1–3. This connection reflects the prior local structure of character "王" in a. During structural matching between strokes $\mathbf{o}_1, \mathbf{o}_2, \mathbf{o}_3, \mathbf{o}_4$ and labels 1–4, we will assign label 4 to stroke $\mathbf{o}_4$ because the relationship between $\mathbf{o}_4$ and $\mathbf{o}_1, \mathbf{o}_2, \mathbf{o}_3$ is the same with the relationship between label 4 and labels 1–3.

The prior clique potentials reflect the prior local structure in the labeling space. We design the single-label prior clique potential for penalizing null labels in Eq. (6.12), and design the pair-label prior clique potential for connected labels in Eq. (6.14).

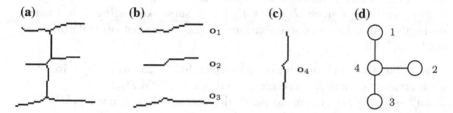

**Fig. 6.15** The connection between labels is prior knowledge about the character structure. In (**d**), four connected labels can represent the structure of character 王 in (**a**), which is constituted by strokes $\mathbf{o}_1, \mathbf{o}_2, \mathbf{o}_3, \mathbf{o}_4$ in (**b**) and (**c**)

In principle, we can design clique potentials arbitrarily if they decrease the values of the energy with an increase of matching degree between input strokes and MRF-based character model [32]. To differentiate characters that are very similar in shape with a small difference as shown in Fig. 6.12, we have to carefully assign weights to some clique potentials that emphasize the subtle structural differences in similar characters. For example, we have to assign large weights to the clique potential for the dot stroke in the character "玉" to emphasize its difference from "王" as shown in Fig. 6.13. These weights for a pair of ambiguous Chinese characters can be automatically obtained by the neural network learning algorithm proposed in [33]. Therefore, after structural matching, we can build a database for ambiguous character pairs, and then differentiate them further by the weighted clique potentials. Alternatively, we can manually set the weights to the clique potentials based on prior knowledge, which can be done during the initialization process.

### 6.4.1.4  MRF-based Stroke Extraction

Extracting reliable strokes is an essential prerequisite to modeling stroke relationships. Before stroke extraction, we normalize the slant and moment with aspect ratio preserved for character images [34]. To reduce the variation of stroke width, we perform EDT-based thinning to the input characters, which can recover the jam-packed holes and remove loosely touching strokes [35]. Then, we extract the endpoints and intersection points from the character skeleton. Meanwhile, we trace the consecutive pixels connecting these points called substrokes, and remove spurious substrokes whose lengths are short [5, 15]. Finally, we break each substroke at high curvature points by corner detection [36].

As shown in Fig. 6.16, the task of stroke extraction is to concatenate substrokes to form reliable strokes, which is difficult because of complicated character shapes and ambiguous intersection regions due to the thinning process. However, the MRF-based character model can provide the global information to concatenate substrokes into reliable strokes. Because of the lack of structural information between substrokes, we can only use the single-site likelihood clique potential (6.8) to extract the candidate strokes that have the minimum likelihood energy to the label $j$. Some other model-based stroke extraction methods can be found in [15, 18].

First, we build a graph $\mathscr{G} = (\mathbf{O}, E)$ for all substrokes, where $E$ is a matrix storing the information of connectable substrokes $\mathbf{o}_i$ and $\mathbf{o}_{i'}$ that satisfy the following conditions [18]:

1. $\mathbf{o}_i$ and $\mathbf{o}_{i'}$ share the same intersection region like $\mathbf{o}_2$ and $\mathbf{o}_3$ in Fig. 6.16, or the distance between their endpoints is less than a threshold $TH_1$.
2. $\|\mathbf{o}_i^D - \mathbf{o}_{i'}^D\| \geq TH_2$, where the intersection region is the origin in Fig. 6.7.

The first condition ensures that two substrokes are from a continuous straight line, and the second condition checks the linearity of two substrokes. For each substroke, we use 2D Gabor filters to extract the cyclic directional observation $\mathbf{o}^D$. For the position and length observations, we use center coordinates $(x, y)$ and the

| labels | substrokes | | | | | | |
|---|---|---|---|---|---|---|---|
| 1 | 2 <br> (3.69) | 2 − 9 <br> (2.23) | 2 − 9 − 7 <br> (1.20) | 2 − 9 − 7 − 8 <br> (1.00) | 2 − 9 − 7 − 8 − 4 <br> (0.29) | ⇒ | CLOSED |
| 2 | 1 <br> (0.64) | 1 − 3 <br> (0.12) | | | | ⇒ | CLOSED |
| 3 | 13 <br> (0.42) | | | | | ⇒ | CLOSED |
| 4 | 11 <br> (0.31) | | | | | ⇒ | CLOSED |
| 5 | 6 <br> (1.71) | 6 − 7 <br> (1.20) | 6 − 7 − 10 <br> (0.35) | | | ⇒ | CLOSED |
| 6 | 5 <br> (1.43) | 5 − 7 <br> (1.18) | 5 − 7 − 9 <br> (0.71) | | | ⇒ | CLOSED |
| 7 | 12 <br> (0.09) | | | | | ⇒ | CLOSED |

**Fig. 6.16** We can build a graph $\mathcal{G} = (\mathbf{O}, E)$ for all substrokes, where $E$ contains collinear information for connectable substrokes. Algorithm 6.17 finds possible concatenations of substrokes that have the minimum likelihood energy to the labels of the MRF. The bracket numbers are the corresponding single-site likelihood potential for each concatenation. The extracted candidate strokes are stored in the CLOSED set

number of pixels of each substroke, both of which are normalized with respect to the character size. As a result, the observation vector $\mathbf{o}_i = [\mathbf{o}_i^D, \mathbf{o}_i^P, \mathbf{o}_i^L]'$ has the dimensionality $d = 11$. We have developed a MRF-based stroke extraction as shown in Fig. 6.17, which searches candidate strokes for each label $j$ of the MRF-based character model. The OPEN set stores the initial substrokes, and the CLOSED set stores all best concatenations of substrokes that minimize the single-site likelihood clique potential (6.8). We denote the number of elements in the OPEN set by |OPEN|. The thresholds $TH_3$ and $TH_4$ control the distance of position and direction from the initial substrokes to the following mean vector of the label $j$,

$$\boldsymbol{\mu}_j = \sum_{m=1}^{M_s} w_{jm} \boldsymbol{\mu}_{jm}. \tag{6.56}$$

Based on the initial substrokes in the OPEN set, Algorithm 6.17 will search all possible connectable substrokes at each iteration, and concatenate them into a new substroke by

---

**input**    : $OPEN, CLOSED, \mathcal{G} = (\mathbf{O}, E), \lambda.$
**output**   : $CLOSED.$
**initialize**: $OPEN, CLOSED \leftarrow \varnothing.$

1  **begin**
2     **for** $j \leftarrow 1$ **to** $J$ **do**
3        **for** $i \leftarrow 1$ **to** $I$ **do**
4           **if** $\|\mathbf{o}_i^P - \boldsymbol{\mu}_j^P\| \leq TH_3,\ \|\mathbf{o}_i^D - \boldsymbol{\mu}_j^D\| \leq TH_4$ **then**
5              $OPEN \leftarrow \mathbf{o}_i;$
6           **end**
7        **end**
8        **for** $i \leftarrow 1$ **to** $|OPEN|$ **do**
9           $\mathbf{o}_{new} \leftarrow \mathbf{o}_i \in OPEN;$
10          **repeat**
11             $\mathbf{o}_{old} \leftarrow \mathbf{o}_{new};$
12             $\mathbf{o}_{new} \leftarrow \mathbf{o}_{old}, \mathbf{o}_{i'}, E;$
13          **until**
               $V_{C_1}(\mathbf{o}_{new}|s_i = j, \lambda) - V_{C_1}(\mathbf{o}_{old}|s_i = j, \lambda) > TH_5$
               ;
14          $CLOSED \leftarrow \mathbf{o}_{old};$
15       **end**
16    **end**
17 **end**

**Fig. 6.17**  The stroke extraction algorithm

$$\mathbf{o}_{new}^P = \frac{\mathbf{o}_i^P \mathbf{o}_i^L + \mathbf{o}_{i'}^P \mathbf{o}_{i'}^L}{\mathbf{o}_i^L + \mathbf{o}_{i'}^L}, \tag{6.57}$$

$$\mathbf{o}_{new}^D = \frac{\mathbf{o}_i^D \mathbf{o}_i^L + \mathbf{o}_{i'}^D \mathbf{o}_{i'}^L}{\mathbf{o}_i^L + \mathbf{o}_{i'}^L}, \tag{6.58}$$

$$\mathbf{o}_{new}^L = \mathbf{o}_i^L + \mathbf{o}_{i'}^L. \tag{6.59}$$

It continues to concatenate this new substroke with possible connectable substrokes until the single-site likelihood potential (6.8) does not decrease further, which is controlled by the threshold $TH_5$. For each label $j$, Fig. 6.17 may extract multiple candidate strokes, and a null label indicates that no candidate strokes are extracted. Finally, we store all candidate strokes $\mathbf{o}_i$ for each label $j$ in the CLOSED set, and remove repetitive ones. We will use the candidate strokes in the CLOSED set and the remaining substrokes as input strokes for structural matching by Fig. 6.2. After structural matching, each label of the MRF will be assigned to one of candidate strokes extracted for it. Therefore, we assign the initial labeling $f_j(i) = 1$ to all candidate strokes $\mathbf{o}_i$ for label $j$, and $f_j(i) = 0$ to other strokes. Figure 6.18 shows some results of the MRF-based stroke extraction, where only the candidate strokes having minimum likelihood energy for each label are shown. After stroke extraction, we will be able to get the neighborhood information of strokes based on their connection.

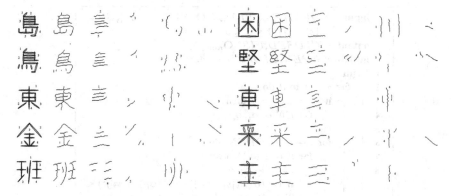

**Fig. 6.18** MRF-based stroke extraction. The first and seventh columns are the corresponding MRF-based character models

## 6.4.2 Handwritten Chinese Character Recognition (HCCR)

An HCCR system based on stroke analysis usually includes five components: a handwritten Chinese character database, character models, language models, the stroke extraction algorithm, and the structural matching algorithm, in which the character models are crucial parts in the HCCR system. Figure 6.19 illustrates the hierarchical structure of the HCCR system based on MRF-based character models. The MRF is the bottom layer that describes the stroke relationships in the stroke layer. The structural matching algorithm like relaxation labeling bridges these two layers. The character layer is the top layer that determines the character relationships by language models [26]. The stroke extraction algorithm bridges the character and stroke

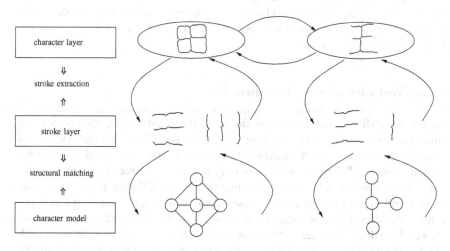

**Fig. 6.19** The hierarchical structure of the HCCR system

$$
\begin{array}{l}
\textbf{input} \quad : OPEN, CLOSED, \mathcal{N}, \mathbf{O} = \{\mathbf{o}_1, \ldots, \mathbf{o}_I\}, \mathcal{S}^* = \\
\qquad\qquad \{s_i^* = j\}, 1 \le j \le J. \\
\textbf{output} \quad : CLOSED, U(\mathcal{S}^* | \mathbf{O}_{new}, \lambda_\omega). \\
\textbf{initialize}: OPEN, CLOSED \leftarrow \varnothing.
\end{array}
$$

1  **begin**
2     **for** $j \leftarrow 1$ **to** $J$ **do**
3        **if** $s_i^* = j, s_{i'}^* = j$ **then**
4           $OPEN \leftarrow \mathbf{o}_i, \mathbf{o}_{i'}$;
5        **end**
6        **for** $i \leftarrow 1$ **to** $|OPEN|$ **do**
7           **for** $i' \leftarrow 1$ **to** $|\mathcal{N}_i|$ **do**
8              $\mathbf{o}_{new} \leftarrow \mathbf{o}_i, \mathbf{o}_{i'}$;
9              **if** $V_{\mathcal{C}_1}(\mathbf{o}_{new} | s_i = j') - V_{\mathcal{C}_1}(\mathbf{o}_{i'} | s_{i'} = j') <$
                $TH_5, V_{\mathcal{C}_1}(\mathbf{o}_{new} | s_i = j') - V_{\mathcal{C}_1}(\mathbf{o}_i | s_i = j) <$
                $TH_5$ **then**
10                $CLOSED \leftarrow \mathbf{o}_{new}$;
11             **end**
12          **end**
13       **end**
14    **end**
15    $U(\mathcal{S}^* | \mathbf{O}_{new}, \lambda_\omega) \leftarrow CLOSED$;
16 **end**

**Fig. 6.20** The stroke concatenation algorithm

layers. To enhance the performance of the HCCR system, we have to improve character models, language models, stroke extraction algorithm, and structural matching algorithm respectively. In this section, we focus on the effectiveness of MRF-based character model, so we do not use language models in the HCCR system. In previous sections, we have presented the RL algorithm, the MRF-based character models, and the MRF-based stroke extraction algorithm. Here, we will propose the model-based broken stroke concatenation as well as learning and recognition algorithms for the MRF-based character models.

### 6.4.2.1 Model-Based Stroke Concatenation

Although the MRF-based stroke extraction algorithm is able to produce most reliable strokes, it still has errors due to thinning distortions as well as complicated shapes of Chinese characters. When the variation of the stroke curvature exceeds a threshold, broken strokes will happen, which not only increase the uncertainty of character structures, but also deteriorate the performance of the HCCR system. To concatenate broken strokes, we propose a model-based stroke concatenation as shown in Algorithm 6.20. After structural matching, we may assign the same label to more than one stroke. We will check if these strokes are neighbors that can be concatenated each other in the neighborhood system $\mathcal{N}$. We concatenate them by Eqs. (6.57)–(6.59), when the $\mathbf{o}_{\text{new}}$ has the lower single-site likelihood clique potential (6.8) than those of previous strokes. Otherwise, we assign the label to the stroke having the lowest likelihood potential, and concatenate another stroke with its connectable neighbors only if the $\mathbf{o}_{\text{new}}$ has the lower likelihood clique potential than those of previous ones. This process repeats until all strokes with the same label have been checked. The aim of the model-based stroke concatenation is to lower the posterior energy (6.5).

### 6.4.2.2  Learning and Recognition Algorithms

The learning process includes three steps: setting up MRF prototypes, initializing MRFs parameters, and Baum-Welch learning MRFs parameters.

First, we set up MRF prototypes for each class of characters using the observation $\mathbf{O}^* = \{\mathbf{o}_1^*, \ldots, \mathbf{o}_I^*\}$ from standard characters of ETL-9B as shown in Fig. 6.22, where the number of sites $I$ of standard characters is equal to the number of labels $J$ of MRF for each class. In this case, the initial mean vectors $\boldsymbol{\mu}_j$ and $\boldsymbol{\mu}_{jj'}$ in Eqs. (6.8) and (6.10) are the unary and binary features $\mathbf{o}_i$ and $\mathbf{o}_{ii'}$ respectively. The initial covariance matrix $\boldsymbol{\Sigma}_j$ and $\boldsymbol{\Sigma}_{jj'}$ in Eqs. (6.8) and (6.10) are set as the diagonal matrix $\text{diag}(\sigma_1^2, \ldots, \sigma_{11}^2)$ due to statistical independence. The initial conditional probability $a_{jj'} = 1$ if the observation $\mathbf{o}_i$ is labeled with $j$ and its neighbor $\mathbf{o}_{i'}$ is labeled with $j'$, otherwise $a_{jj'} = 0$. We can see that all initial information about character structures are automatically obtained from observations of the standard characters, which means that the MRF-based character models are initialized by the standard characters.

Second, for each training character image, we extract strokes and obtain the observation set $\mathbf{O}$. Suppose a set of training observations $\mathbf{O}^r$, $1 \le r \le R$, is used to estimate the parameters of an MRF-based character model with $M_s$ mixture components. We use the RL algorithm to assign labels to training observations $\mathbf{O}^r$, $1 \le r \le R$. The best labeling configuration $\mathcal{S}^*$ implies an alignment of observations $\{\mathbf{o}_1, \ldots, \mathbf{o}_I\}$ with labels. We use the k-means algorithm to further cluster the observations within the same label into different mixture components [37]. As a consequence, every observation is associated with a single unique mixture component. This association can be represented by the indicator function,

$$\Theta_{jm}^r(i) = \begin{cases} 1, & \text{if } \mathbf{o}_i^r \text{ is with the } m\text{th mixture component of label } j, \\ 0, & \text{otherwise.} \end{cases} \quad (6.60)$$

Therefore, the mean vector, covariance matrix, mixture weight, and $a_{jj'}$ of the single-site likelihood clique potential can be estimated via simple averages as shown in Fig. 6.3. The estimation of pair-site likelihood clique potential is almost the same except that we use the binary features $\mathbf{o}_{ii'}$ and indicator function,

```
input   : BW, λ_ω, ω = 1, ..., C.
output  : ω*.
1 begin
2   for ω ← 1 to C do
3       O ← strokeextraction(BW, λ_ω);
4       U(S*|O, λ_ω) ← relaxationlabeling(O, λ_ω);
5       U(S*|O_new, λ_ω) ←
            strokeconcatenation(O, S*);
6   end
7   ω* ← arg min_ω U(S*|O_new, λ_ω) ;
8 end
```

**Fig. 6.21** The recognition algorithm

$$\Theta^r_{jj'm}(i) = \begin{cases} 1, & \text{if } \mathbf{o}^r_i, \mathbf{o}^r_{i'} \text{ is with the } m\text{th mixture component of label } j, j', \\ 0, & \text{otherwise.} \end{cases}$$

(6.61)

Finally, we use the Baum-Welch algorithm to refine all parameters of MRFs according to the ML criterion [37]. Given a set of training observations, the Baum-Welch algorithm can iteratively and automatically adjust parameters $\mu$ and $\Sigma$ in the $m$th mixture component of the MRF. Once again, we use the RL algorithm without the winner-take-all strategy. After iterations, the RL terminates and associates each observation $\mathbf{o}_i$ with the $j$th label and $m$th mixture by the posterior probability $f_{jm}(i) \in [0, 1]$, which represents the $\mathbf{o}_i$'s contribution to computing the maximum likelihood parameter values for label $j$ and mixture $m$. In other words, rather than assigning a specific label to each site, we assign the label to each site in proportion to the posterior probability $f^r_{jm}(i)$. The process of the Baum-Welch learning is shown in Fig. 6.4, except here we use $f_{jm}(i)$ as the updating weight. To describe uncertainty of character structures, we may set up more than one prototype for each class of characters. The use of multiple prototypes adds no additional complexity to the learning algorithm, which can be repeated for each prototype learning. To get accurate character models, a large amount of training observations is needed. When the number of training observations is small, certain mixture components will have very few associated training observations, so the variances or the corresponding mixture weight will be very small. In such cases, the mixture component is deleted, provided that at least one component in that label is left. Because of limited training observations, the covariance matrix may be singular and irreversible. So the Baum-Welch algorithm updates only mean vectors, and leaves the covariance matrix unchanged in this case.

In the MAP-MRF framework, the recognition as shown in Fig. 6.21 is formulated to find the MRF-based character model $\omega^*$ that can minimize the energy $U(\mathcal{S}^*|\mathbf{O}, \lambda_\omega)$ for input character image $BW$.

**Fig. 6.22**   The *first line* shows samples from ETL-9B. The *second line* shows the standard characters in ETL-9B. The *third line* shows samples from Hanja1 database. The *fourth line* shows samples from Hanja2 database

### 6.4.3  Experimental Results

#### 6.4.3.1  Handwritten Chinese Character Databases

ETL-9B is a public database of handwritten Chinese characters [38], which includes 2, 965 classes of handprinted Chinese characters, and 71 classes of Japanese characters. The characters have been written by 4000 writers with 200 samples for each class. Besides, each class has one sample of well-written standard character with clear stroke trajectory. All samples are binary bitmap files with $63 \times 64$ pixels resolution. Some samples are shown in Fig. 6.22.

The KAIST includes Hanja1 and Hanja2 databases that are publicly available [15, 18]. The Hanja1 database has 783 classes with 200 samples for each class. The Hanja2 database has 1309 samples from real documents only for test purpose. The image quality of Hanja1 database is good, but Hanja2 database is bad. Some samples are shown in Fig. 6.22.

#### 6.4.3.2  Similar Character Recognition on ETL-9B Database

We carried out experiments on similar character recognition of ETL-9B. The character images in ETL-9B have no noise, and the characters are neatly written. So far, the highest recognition rate 99.42 % was reported in [23], which used directional element feature and asymmetric Mahalanobis distance for fine classification. However, the major reason of misclassification is the ambiguities of shapes in similar characters [18, 25, 31, 33]. Therefore, to investigate the effectiveness of statistical-structural character modeling by MRFs, we selected 50 pairs of highly similar Chinese characters in the database ETL-9B as the recognition vocabulary in our experiments as shown in Fig. 6.23. The vocabulary was constituted by choosing the most confusable

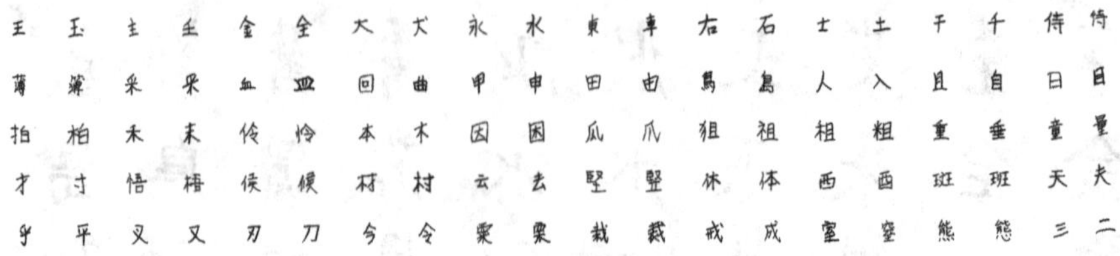

王 玉 主 生 金 全 大 犬 永 水 東 隼 右 石 土 土 于 千 侍 侍

莆 笋 釆 釆 血 皿 回 曲 甲 申 田 由 馬 鳥 人 入 且 自 日 日

拍 柏 未 末 佮 伶 本 木 囚 困 瓜 爪 狙 祖 相 粗 重 垂 童 量

才 寸 悟 梅 侯 候 村 村 云 去 堅 竪 休 体 西 酉 斑 班 天 夫

少 平 又 又 刀 刀 兮 兮 栗 栗 栽 戴 戒 成 窒 窒 熊 態 三 二

**Fig. 6.23** A vocabulary of 50 pairs of highly similar Chinese characters

**Table 6.1** Recognition rate of similar characters

| Database | Xiong et al. [31] (%) | Proposed (%) | Emphasize clique potentials (%) |
|---|---|---|---|
| ETL-9B | 95.5 | 96.9 | 98.5 |

pairs reported in [31]. Each character had 200 samples of which 150 were used for training and the remaining 50 for recognition.

We used two mixtures in the MRF-based character model because of limited training samples. We manually set the weights to clique potentials in order to emphasize the subtle difference between similar characters, which can be also done automatically [33]. Table 6.1 shows the comparison with that of Xiong et al. [31]. The average recognition rate of our HCCR system before setting weights of clique potentials was 96.9 %, which is higher than the best result 95.5 % reported in [31]. We used the same amount of training samples as [31], but [31] used a different and publicly unavailable database for both training and recognition. After setting weights to clique potentials marked by dark in Fig. 6.24, the HCCR system successfully correct those misclassified samples as shown in Fig. 6.24. The recognition rate increases to 98.5 %, which is three percent higher than that of Xiong et al. [31]. The experimental results demonstrates that the MRF-based character models can represent the fine detail of character structures to differentiate similar characters.

### 6.4.3.3 Character Recognition on KAIST Database

We also evaluated the performance of MRF-based character models on the KAIST Hanja1 and Hanja2 databases. Some typical samples in Hanja1 and Hanja2 are illustrated in Fig. 6.22. The baseline recognizor used the odd number of samples of every class for training, and the first ten samples of the even class for the test on the Hanja1 database [18]. By handling degraded region, the baseline recognition rate was 98.45 % [18]. The Hanja2 database was only used for test with recognition rate 83.14 %. Based on baseline recognisor, a binary classifier was proposed to further differentiate similar characters, which improved the overall recognition rate from 98.45 to 99.46 % on Hanja1 database. In our experiment, we selected 783 classes

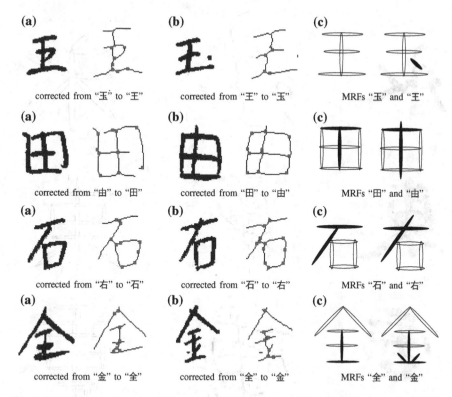

**Fig. 6.24** Examples of corrected samples. The *dark clique* potentials of MRFs are emphasized

of characters as recognition vocabulary from Hanja1 database. For each class, we used ten samples of the even number for the test and the remaining 190 samples for training. To evaluate the MRF-based character model for cursive Chinese characters, we also used the samples from the Hanja2 database for the test. To compare with the baseline recognizer, we did not specially differentiate the similar characters as in [33].

Figure 6.25 shows the MRF-based stroke extraction and structural matching results. The first column shows the input character images. The second column shows the slant and moment normalization of the character skeleton. The third column shows the MRF-based character models, where the labels are numbered. On the second column, Algorithm 6.2 assigns the labels to extracted strokes. Table 6.2 shows the comparison with other recognizers on Hanja1 and Hanja2 databases [15, 18]. Kim and Kim used the first 100 odd number of samples in Hanja1 for training, and the first 10 samples of even number for recognition [18]. Liu et al. used the first 80 odd number of samples in Hanja1 for training, and the first 20 samples of even number for recognition [15]. The recognition rate of MRF-based HCCR system was 0.98 and 0.42 % higher than those reported in [15, 18] respectively, though we used more training samples. The recognition rate on Hanja2 database was 0.93 %

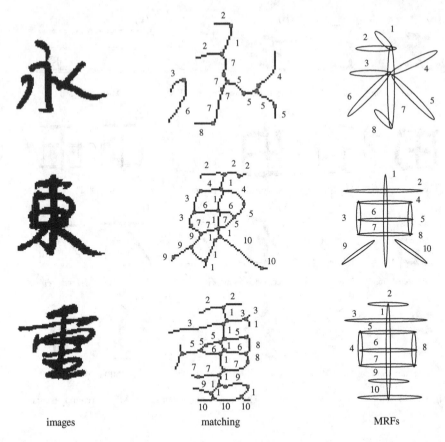

images                         matching                         MRFs

**Fig. 6.25** MRF-based stroke extraction and matching results

**Table 6.2** Recognition rate comparison on KAIST database

| Database | Liu et al. [15] (%) | Kim and Kim [33] (%) | Kim and Kim [18] (%) | Proposed (%) |
|----------|---------------------|----------------------|----------------------|--------------|
| Hanja1   | 97.89               | 99.46                | 98.45                | 98.87        |
| Hanja2   | –                   | –                    | 83.14                | 82.21        |

lower because the character images in Hanja2 have many degraded regions that deteriorate the stroke extraction. However, the recognition rate was still 0.56 % higher than the intermediate result 81.65 % in [18]. Some misclassified samples are shown in Fig. 6.26. Two reasons lead to the misclassification: (1) character image degradation, and (2) similar characters [18]. The first may be solved by proper preprocessing and pseudostrokes [18]. The second can be solved by setting weights to the clique potentials to emphasize subtle difference between similar characters [33].

The MRF has been demonstrated effective for handwritten Chinese character recognition by modeling stroke relationships. We also applied the IT2 FMRF to

"主" → "王"              "壬" → "王"              "己" → "乙"              "太" → "大"

**Fig. 6.26**  Misclassified samples

**Fig. 6.27**  Similar Chinese
characters

similar character recognition on ETL-9B and KAIST Hanja1 databases [18, 38]. Without loss of generality, we selected only 16 highly similar characters as recognition vocabulary as shown in Fig. 6.27. Some of them have little difference and share almost the same structures or substructures.

For comparison, we used two mixtures in both MRFs and IT2 FMRFs, and defined the neighborhood system based on the stroke connection of Chinese characters. We classified unknown test data to the IT2 FMRF having the minimum energy in Eq. (6.26). Because $U_{\tilde{\lambda}}(\tilde{\mathscr{S}}^*|\mathbf{O})$ is an interval set, we have three choices such as the left-end point, central, and right-end point to classify data. If the uncertainty factor $k = 0$, those three points reduce to one point, and the energy is the same with that of the MRF. The bigger $k$, the larger range of interval set $U_{\tilde{\lambda}}(\tilde{\mathscr{S}}^*|\mathbf{O})$, which may have higher possibility to contain the best labeling energy for observations. In practice, we set $k \in (0.4, 1)$ to account for all possibilities of posterior energy. We took the center value and the range of the prior belief $\tilde{v}_1$ and $\tilde{v}_2$ in Eqs. (6.12) and (6.14) based on our experience. In this section, we ranked interval sets $U_{\tilde{\lambda}}(\tilde{\mathscr{S}}^*|\mathbf{O})$ directly without defuzzification as mentioned in Sect. 4.3, which can get better results without loss of information.

To compare the classification ability between MRFs and IT2 FMRFs, we trained them by 100 samples and tested them by the rest 100 samples in ETL-9B and Hanja1 databases for each character class. Table 6.3 shows the results on the two databases.

**Table 6.3**  Classification ability

| Training / Test databases | MRFs (%) | IT2 FMRFs (%) |
| --- | --- | --- |
| ETL-9B / ETL-9B | 94.81 | 96.57 |
| Hanja1 / Hanja1 | 93.97 | 96.45 |

**Table 6.4**  Generalization ability

| Training / Test databases | MRFs (%) | IT2 FMRFs (%) |
| --- | --- | --- |
| ETL-9B / Hanja1 | 86.30 | 91.56 |
| Hanja1 / ETL-9B | 89.92 | 92.84 |

The IT2 FMRF has 2.21 % higher classification rate than the MRF on average. To compare the generalization ability, we trained IT2 FMRFs and MRFs by 200 samples from ETL-9B for each character, and tested them by 200 samples from Hanja1 database, and vice versa. Table 6.4 shows that IT2 FMRFs degrade only 4.31 % whereas MRF degrade 6.28 % on average in terms of classification rate, which demonstrate the IT2 FMRF has a better generalization ability than the MRF.

## 6.5  Summary

The great success achieved in pattern recognition, image processing, and computer vision in the passing decades has been largely due to the recognition and exploration local statistical dependencies existing universally in patterns, images, and video frames. MRFs have been introduced to these research disciplines by pioneer researchers such as Geman and Geman, Chellappa and Jain, Kindermann and Snell, and Li [9, 28, 29, 39]. Within the MRF framework, statistical interaction among labels at adjacent sites in a pattern or image are reflected by two fundamental concepts: *neighborhood system* $\mathcal{N}$ and *clique potentials* $V_c$. Therefore, the pattern's structure can be represented statistically by the MRF. By use of the equivalence between MRFs and GRFs due to the Hammersley-Clifford theorem [8], the best global configuration can be reached by computing MAP of the MRF. This MAP-MRF framework has found an extremely broad spectrum of applications of interest to pattern recognition, image processing, and computer vision summarized in [9].

In HMMs, we often compute the log-likelihood and probability to prevent underflow rather than direct computation of them. As we will see, the log-likelihood and probability are actually clique potentials introduced in MRFs. Actually, HMMs are a class of the simplest MRFs where HMMs' neighborhood systems $\mathcal{N}_i$ only contain the previous time $i - 1$. Unlike HMMs, the inference of MRFs is not easy, so that we use the RL algorithm to solve it approximately [40, 41]. We have found that the RL algorithm is actually a generalization of the forward-backward and Viterbi algorithms. However, such a generalization is not strict because the RL considers only the combinatorial problem within the neighborhood system, while the forward-backward and Viterbi algorithms consider the global combinatorial problem due to the simple neighborhood system in HMMs.

From graphical models point of view, the prior constraints in the labeling config-uration implies a graph with a certain structure, where the labels are nodes and the relationships between labels are edges, as for instance, the transition probability $a_{jj'}$ is the weight for the edge between label $j$ and $j'$ in HMMs; the prior clique potential $V_{\mathscr{C}_2}(j, j'|\lambda)$ is the weight for the edge between label $j$ and $j'$ in MRFs. The sites compose another graph, where the site $i$ have only edges with its neighboring sites $\mathscr{N}_i$. Therefore, the labeling problem can be viewed as the process of graph matching between sites and HMMs or MRFs, which can be solved by the forward-backward or RL algorithms according to the maximum posterior probability or minimum posterior energy criterion.

In practice, we avoid evaluating the partition function $Z^{-1}$ in MRFs [9], which actually circumvents the global combinatorial problem. Theoretically speaking, learning parameters of MRFs from data must evaluate the partition function [9]. However, such an approximation usually provides satisfactory results in many appli-cations, so we generally ignore the partition function in the design of clique potentials.

MRFs have been widely used for texture analysis and segmentation [9, 28, 29, 42, 43]. Wei extended edit distance within the MRF referred to as Markov edit distance [44]. Zeng and Liu applied MRFs to handwritten Chinese character modeling and recognition [1, 45]. Other applications such as object recognition, face detection, and recognition can be found in [9, 46].

The stroke segmentation problem has been formulated within the theoretically well-founded MAP-MRF framework. For simplicity, we use only a second-order neighborhood system. We obtain the continuous segmented strokes by the optimal labeling rather than *ad hoc* rules. The experimental results show that the ambiguous parts due to distortions can be accurately detected. Furthermore, the computational complexity is not high because of the correct initial labeling and the fewer number of sites by thinning process. However, the result is dependent on the directional features of pixel, which is not always fall in four predefined directions. Therefore, we prefer model-based stroke extraction in practice.

We have proposed a systematic strategy for statistical character structure modeling within the MAP-MRF framework. This statistical-structural strategy can effectively represent structural patterns like Chinese characters in terms of three issues:

1. The neighborhood system that defines the most important stroke relationships;
2. Clique potentials that measures the distance between the strokes and character models;
3. The structural matching algorithm that finds the best labeling configuration.

Corresponding to the three issues, we have proposed (1) the neighborhood system based on connection of strokes, (2) the clique potentials derived from GMMs and prior knowledge, and (3) the RL algorithm.

Indeed, some other methods modeling stroke relationships can also be considered within the MAP-MRF framework. For example, in [15], (1) the neighborhood system was manually defined by categorizing the strokes into several relational types; (2)

the matching distance between input strokes and models was designed heuristically; and (3) the matching algorithm was a heuristic search strategy. In [18], (1) the neighborhood system was defined by *Kullback-Leibler measure*; (2) the Gaussian joint distribution of strokes and models was used as the matching distance; and (3) the heuristic search algorithm found the optimal matching. Compared with above methods, the MRF-based character model is theoretically well-founded, and has many choices to design the neighborhood system and clique potentials. Especially, the pair-site likelihood clique potential for binary features can encode structural information of the character from training samples, which is different from other character structure modeling methods. Furthermore, the RL algorithm is a fast parallel minimizer of MRFs energy function for structural matching, which may be more robust than heuristic search strategies in [15, 18].

We have also proposed a new type of stroke features based on two-dimensional Gabor filters, in which the direction of the stroke has a cyclic representation. After normalization, the direction, position, and length of the stroke can be put into a vector with dimensionality $d = 11$. One advantage of this feature vector is that we do not need additional coefficients to balance the effect of direction, position, and length of the stroke in calculating the matching distance [15].

The model-based stroke extraction is able to extract reliable stroke candidates for structural matching [15, 18]. To this end, we have proposed the MRF-based stroke extraction method that extracts the candidate strokes having the minimum single-site likelihood clique potential for each label. Furthermore, we have proposed model-based stroke concatenation after structural matching. We concatenate two strokes if they can be concatenated and assigned with the same label, and the concatenated stroke has a lower single-site likelihood clique potential than those of the previous ones. Since the MRF-based character model encodes the global stroke information from training samples, incorporating stroke extraction and concatenation in the MAP-MRF framework is indeed promising.

Within the MAP-MRF framework, we have built an HCCR system that systematically learns parameters of MRF-based character model from training samples. The recognition rate of similar characters on the ETL-9B database was 3.0 % higher than that reported in [31], and the recognition rate on the Hanja1 database was 0.42 and 0.98 % higher than those reported in [15, 18], which demonstrates the effectiveness of MRFs for statistical-structural character modeling.

Hidden fuzzy Markov random fields have been used for classical image segmentation [47, 48]. They assume some image pixels are fuzzy and difficult to segment so that fuzzy fields can characterize theses pixels. However, the T2 FMRF assumes the sites are crisp (image pixels or regions), but the labels assigned to sites are fuzzy, which are evaluated by fuzzy clique potentials. By the equivalence between MRFs and GRFs [8], the uncertainties in the fuzzy likelihood energy and prior energy are translated into fuzzy posterior energy based on the T2 fuzzy Bayesian decision theory [49]. We also compare the results of using IT2 FMRFs and MRFs for similar handwritten Chinese character recognition. The experimental results show a better generalization ability of the IT2 FMRF.

# References

1. Zeng, J., Liu, Z.Q.: Markov random fields for handwritten Chinese character recognition. In: 8th International Conference on Document Analysis and Recognition, pp. 101–105 (2005)
2. Su, Y.M., Wang, J.F.: A novel stroke extraction method for chinese character using gabor filters. Pattern Recognit. **36**(3), 635–647 (2003)
3. Su, Y.M., Wang, J.F.: Decomposing Chinese characters into stroke segments using SOGD filters and orientation normalization. In: 17th International Conference on Pattern Recognition, vol. 2, pp. 351–354 (2004)
4. Liu, K., Huang, Y.S., Suen, C.Y.: Robust stroke segmentation method for handwritten chinese character recognition. In: 4th Interantional Conference on Document Analysis and Recognition, vol. 1, pp. 211–215 (1997)
5. Lin, F., Tang, X.: Off-line handwritten chinese character stroke extraction. In: 16th Interantional Conference on Pattern Recognition, vol. 3, pp. 249–252 (2002)
6. Cao, R., Tan, C.L.: A model of stroke extraction from Chinese character images. In: 15th Interantional Conference on Pattern Recognition, vol. 4, pp. 368–371 (2000)
7. Fan, K.C., Wu, W.H.: A run-length coding based approach to stroke extraction of chinese characters. In: 15th Interantional Conference on Pattern Recognition, vol. 2, pp. 565–568 (2000)
8. Hammersley, J.M., Clifford, P.: Markov Field on Finite Graphs and Lattices. unpublished (1971)
9. Li, S.Z.: Markov Random Field Modeling in Image Analysis. Springer, New York (2001)
10. Movellan, J.R.: Tutorial on Gabor Filters (2002). http://mplab.ucsd.edu/tutorials/pdfs/Gabor.pdf
11. Buse, R., Liu, Z.Q.: Using gabor filters to measure the physical parameters of line. Pattern Recognit. **29**(4), 615–625 (1996)
12. Buse, R., Liu, Z.Q., Caelli, T.: A structural and relational approach to handwritten word recognition. IEEE Trans. Syst. Man Cybern. **B 27**(5), 847–861 (1997)
13. Lee, T.S.: Image representation using 2D gabor wavelets. IEEE Trans. Pattern Anal. Mach. Intell. **18**(10), 1–13 (1996)
14. Hamamoto, Y., et al.: A gabor filter-based method for recognizing handwriting numerals. Pattern Recognit. **31**(4), 395–400 (1998)
15. Liu, C.L., Kim, I.J., Kim, J.H.: Model-based stroke extraction and matching for handwritten chinese character recognition. Pattern Recognit. **34**(12), 2339–2352 (2001)
16. Liu, Z.Q., Cai, J., Buse, R.: Handwriting Recognition: Soft Computing and Probabilistic Approaches. Springer, Berlin (2003)
17. Wang, X., Mohanty, N., McCallum, A.: Group and topic discovery from relations and text. In: LinkKDD, pp. 28–35 (2005)
18. Kim, I.J., Kim, J.H.: Statistical character structure modeling and its application to handwriting chinese character recognition. IEEE Trans. Pattern Anal. Mach. Intell. **25**(11), 1422–1436 (2003)
19. Lin, J.R., Chen, C.F.: Stroke extraction for chinese characters using a trend-followed transcribing technique. Pattern Recognit. **29**(11), 1789–1805 (1996)
20. Liu, C.L., Jaeger, S., Nakagawa, M.: Online recognition of chinese characters: the state-of-the-art. IEEE Trans. Pattern Anal. Mach. Intell. **26**(2), 198–213 (2004)
21. Huang, X., Gu, J., Wu, Y.: A constrained approach to multifont Chinese character recognition. IEEE Trans. Pattern Anal. Mach. Intell. **15**(8), 838–843 (1993)
22. Kang, K.W., Kim, J.H.: Utilization of hierarchical, stochastic relationship modeling for Hangul character recognition. IEEE Trans. Pattern Anal. Mach. Intell. **26**(9), 1185–1196 (2004)
23. Kato, N., Suzuki, M., Omachi, S., Aso, H., Nemoto, Y.: A handwritten character recognition system using directional element feature and asymmetric mahalanobis distance. IEEE Trans. Pattern Anal. Mach. Intell. **21**(3), 258–262 (1999)
24. Liu, C.L., Nakagawa, M.: Evaluation of prototype learning algorithms for nearest-neighbor classifier in application to handwritten character recognition. Pattern Recognit. **34**(3), 601–615 (2001)

25. Tang, Y.Y., Tu, L.T., Liu, J., Lee, S.W., Lin, W.W.: Off-line recognition of chinese handwriting by multifeature and multilevel classification. IEEE Trans. Pattern Anal. Mach. Intell. **20**(5), 556–561 (1998)
26. Wong, P.K., Chan, C.: Off-line handwritten chinese character recognition as a compound bayes decision problem. IEEE Trans. Pattern Anal. Mach. Intell. **20**(9), 1016–1023 (1998)
27. Kim, H.Y., Kim, J.H.: Hierarchical random graph representation of handwritten characters and its application to hangul recognition. Pattern Recognit. **34**(2), 187–201 (2001)
28. Chellappa, R., Jain, A. (eds.): Markov Random Fields: Theory and Application. Academic Press, Boston (1993)
29. Geman, S., Geman, D.: Stochastic relaxation, Gibbs distribution and the Bayesian restoration of images. IEEE Trans. Pattern Anal. Mach. Intell. **6**(6), 721–741 (1984)
30. Wang, Q., Chi, Z., Feng, D., Zhao, R.: Hidden Markov random field based approach for off-line handwritten chinese character recognition. In: 15th Interantional Conference on Pattern Recognition, vol. 2, pp. 347–350 (2000)
31. Xiong, Y., Huo, Q., Chan, C.: A discrete contextual stochastic model for the offline recognition of handwritten chinese characters. IEEE Trans. Pattern Anal. Mach. Intell. **23**(7), 774–782 (2001)
32. Cai, J., Liu, Z.Q.: Pattern recognition using Markov random field models. Pattern Recognit. **35**(3), 725–733 (2002)
33. Kim, I.J., Kim, J.H.: Pair-wise discrimination based on a stroke importance measure. Pattern Recognit. **35**(10), 2259–2266 (2002)
34. Liu, C.L., Nakashima, K., Sako, H., Fujisawa, H.: Handwritten digit recognition: investigation of normalization and feature extraction techniques. Pattern Recognit. **37**(2), 265–279 (2004)
35. Chang, H.H., Yan, H.: Analysis of stroke structures of handwritten chinese characters. IEEE Trans. Syst. Man Cybern. **B 29**(1), 47–61 (1999)
36. He, X., Yung, N.: Curvature scale space corner detector with adaptive threshold and dynamic region of support. In: Proceedings of 17th Interantional Conference Pattern Recognition, vol. 2, pp. 791–794 (2004)
37. Duda, R.O., Hart, P.E., Stork, D.G.: Pattern Classification, 2nd edn. Wiley, New York (2001)
38. Saito, T., Yamada, H., Yamamoto, K.: On the database ETL9 of handprinted characters in JIS chinese characters and its analysis. Trans. IEICE **J68–D**(4), 757–764 (1985)
39. Kindermann, R., Snell, J.L.: Markov Random Fields and Their Applications. American Mathematical Society, Providence (1980)
40. Li, S.Z., Wang, H., Chan, K.L.: Minimization of MRF energy with relaxation labeling. J. Math. Imaging Vis. **7**(2), 149–161 (1997)
41. Faber, P.: A theoretical framework for relaxation processes in pattern recognition: application to robust nonparametric contour generalization. IEEE Trans. Pattern Anal. Mach. Intell. **25**(8), 1021–1027 (2003)
42. Zhu, S.C.: Statistical modeling and conceptualization of visual patterns. IEEE Trans. Pattern Anal. Mach. Intell. **25**(6), 691–712 (2003)
43. Guo, C.E., Zhu, S.C., Wu, Y.N.: Modeling visual patterns by integrating descriptive and generative methods. Int. J. Comput. Vis. **53**(1), 5–29 (2003)
44. Wei, J.: Markov edit distance. IEEE Trans. Pattern Anal. Mach. Intell. **26**(3), 311–321 (2004)
45. Zeng, J., Liu, Z.Q.: Markov random fields-based statistical character structure modeling for chinese character recognition. IEEE Trans. Pattern Anal. Mach. Intell. **30**(5), 767–780 (2008)
46. Li, S.Z., Jain, A.K. (eds.): Handbook of Face Recognition. Springer, New York (2005)
47. Salzenstein, F., Pieczynski, W.: Parameter estimation in hidden fuzzy Markov random fields and image segmentation. Gr. Models Image Process. **59**(4), 205–220 (1997)
48. Ruan, S., Moretti, B., Fadili, J., Bloyet, D.: Segmentation of magnetic resonance images using fuzzy Markov randomfields. In: Proceedings of the International Conference on Image Processing. **3**, 1051–1054 (2001)
49. Zeng, J.: A topic modeling toolbox using belief propagation. J. Mach. Learn. Res. **13**, 2233–2236 (2012)

# Chapter 7
# Type-2 Fuzzy Topic Models

**Abstract** Latent Dirichlet allocation (LDA) is an important hierarchical Bayesian model for probabilistic topic modeling, which attracts worldwide interests and touches on many important applications in text mining, computer vision and computational biology. We first introduce a novel inference algorithm, called belief propagation (BP), for learning LDA, and then introduce how to speed up BP for fast topic modeling tasks. Following the "bag-of-words" (BOW) representation for video sequences, this chapter also introduces novel type-2 fuzzy topic models (T2 FTM) to recognize human actions. In traditional topic models (TM) for visual recognition, each video sequence is modeled as a "document" composed of spatial–temporal interest points called visual "word". Topic models automatically assign a "topic" label to explain the action category of each word, so that each video sequence becomes a mixture of action topics for recognition. The T2 FTM differs from previous TMs in that it uses type-2 fuzzy sets (T2 FS) to encode the semantic uncertainty of each topic. We ca use the primary membership function (MF) to measure the degree of uncertainty that a document or a visual word belongs to a specific action topic, and use the secondary MF to evaluate the fuzziness of the primary MF itself. In this chapter, we implement two T2 FTMs: (1) interval T2 FTM (IT2 FTM) with all secondary grades equal one, and (2) vertical-slice T2 FTM (VT2 FTM) with unequal secondary grades based on our prior knowledge. To estimate parameters in T2 FTMs, we derive the efficient message-passing algorithm. Experiments on KTH and Weizmann human action data sets demonstrate that T2 FTMs are better than TMs to encode visual word uncertainties for human action recognition.

## 7.1 Latent Dirichlet Allocation

Latent Dirichlet allocation (LDA) [8] is a three-layer hierarchical Bayesian model (HBM) that can infer probabilistic word clusters called topics from the document-word matrix. Variational Bayes (VB) [8] and collapsed Gibbs sampling (GS) [26] have been two commonly used approximate inference methods for learning LDA and its extensions, such as author-topic models (ATM) [51] and relational topic models (RTM) [16]. Other inference methods for probabilistic topic modeling

© Tsinghua University Press, Beijing and Springer-Verlag Berlin Heidelberg 2015     129
J. Zeng and Z.-Q. Liu, *Type-2 Fuzzy Graphical Models for Pattern Recognition*,
Studies in Computational Intelligence 591, DOI 10.1007/978-3-662-44690-4_7

include expectation propagation (EP) [43] and collapsed VB inference (CVB) [58]. The connections and empirical comparisons among these approximate inference methods can be found in [2]. Recently, LDA and HBMs have found many important real-world applications in text mining and computer vision (e.g., tracking historical topics from time-stamped documents [49] and activity perception in crowded and complicated scenes [63]).

This chapter represents the collapsed LDA using a novel factor graph [33], by which the topic modeling problem can be interpreted as a labeling problem. The objective is to assign a set of semantic topic labels to explain the observed nonzero elements in the document-word matrix. The factor graph solves the labeling problem by two important concepts: *neighborhood systems* and *factor functions* [6]. It assigns the best topic labels according to the *maximum a posteriori* (MAP) estimation through maximizing the posterior probability, which is often a prohibitive combinatorial optimization problem in the discrete topic space.

The factor graph is a graphical representation method for both directed models (e.g., hidden Markov models (HMMs) [6, Chap. 13.2.3]) and undirected models (e.g., Markov random fields (MRFs) [6, Chap. 8.4.3]) because factor functions can represent both conditional and joint probabilities. In this chapter, we use the factor graph to represent the *collapsed* LDA by encoding the same joint probability. The basic idea is inspired by the collapsed GS algorithm for LDA [26, 27], which integrates out the multinomial parameters based on the Dirichlet-Multinomial conjugacy and views the Dirichlet hyperparameters as the pseudo topic labels having the same layer with the latent topic labels. In the collapsed hidden variable space, we find that the joint probability of the collapsed LDA can be decomposed as the product of factor functions in the factor graph. This collapsed view has also been discussed within the mean-field framework [3], inspiring the zero-order approximation CVB (CVB0) algorithm [2] for learning LDA. By contrast, the undirected model "harmonium" [64] encodes a different joint probability from LDA and probabilistic latent semantic analysis (PLSA) [29], so that it is a new and viable alternative to these directed models.

The factor graph of the collapsed LDA facilitates the classic loopy belief propagation (BP) algorithm [6, 24, 33] for approximate inference and parameter estimation. By designing the *neighborhood system* and *factor functions*, we may encourage or penalize different local labeling configurations to realize the topic modeling goal. The BP algorithm operates well on the factor graph, and it has the potential to become a generic learning scheme for variants of LDA-based topic models in the collapsed space. For example, we also extend the BP algorithm to learn ATM [51] and RTM [16] based on the factor graph representations. The convergence of BP is not guaranteed on general factor graphs [6], but it often converges and works well in real-world applications. Although learning generic HBMs often has difficulty in estimating parameters and inferring hidden variables due to causal coupling effects, the factor graph as well as the BP-based learning scheme may shed more light on faster and more accurate algorithms for learning generic HBMs (Table. 7.1).

The probabilistic topic modeling task is to assign a set of semantic topic labels, $\mathbf{z} = \{z_{w,d}^k\}$, to explain the observed nonzero elements in the document-word matrix

**Table 7.1** Notations

| | |
|---|---|
| $1 \leq d \leq D$ | Document index |
| $1 \leq w \leq W$ | Word index in vocabulary |
| $1 \leq k \leq K$ | Topic index |
| $1 \leq a \leq A$ | Author index |
| $1 \leq c \leq C$ | Link index |
| $\mathbf{x} = \{x_{w,d}\}$ | Document-word matrix |
| $\mathbf{z} = \{z_{w,d}^k\}$ | Topic labels for words |
| $\mathbf{z}_{-w,d}$ | Labels for $d$ excluding $w$ |
| $\mathbf{z}_{w,-d}$ | Labels for $w$ excluding $d$ |
| $\mathbf{a}_d$ | Coauthors of the document $d$ |
| $\mu_{\cdot,d}(k)$ | $\sum_w x_{w,d}\mu_{w,d}(k)$ |
| $\mu_{w,\cdot}(k)$ | $\sum_d x_{w,d}\mu_{w,d}(k)$ |
| $\theta_d$ | Factor of the document $d$ |
| $\phi_w$ | Factor of the word $w$ |
| $\eta_c$ | Factor of the link $c$ |
| $f(\cdot)$ | Factor functions |
| $\alpha, \beta$ | Dirichlet hyperparameters |

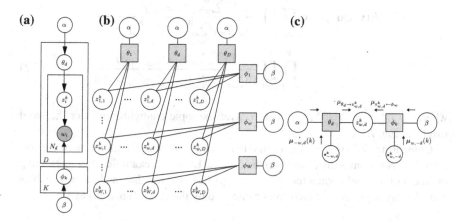

**Fig. 7.1** **a** Three-layer LDA, **b** two-layer factor graph for the collapsed LDA, and **c** message passing

$\mathbf{x} = \{x_{w,d}\}$. The notation $1 \leq k \leq K$ is the topic index, $x_{w,d}$ is the number of word counts at the index $\{w, d\}$, $1 \leq w \leq W$ and $1 \leq d \leq D$ are the word indexes in the vocabulary and the document index in the corpus. Table 7.1 summarizes some important notations.

Figure 7.1a shows the three-layer graphical representation of LDA [8]. The document-specific topic proportion $\theta_d(k)$ generates a topic label $z_{w,d,i}^k \in \{0, 1\}$, $\sum_{k=1}^K z_{w,d,i}^k = 1$, which in turn generates each observed word token $i$ at the index $w$

in the document $d$ based on the topic-specific multinomial distribution $\phi_k(w)$ over the vocabulary words. Both multinomial parameters $\theta_d(k)$ and $\phi_k(w)$ are generated by two Dirichlet distributions with hyperparameters $\alpha$ and $\beta$, respectively. For simplicity, we consider only the smoothed LDA [26] with the fixed symmetric Dirichlet hyperparameters $\alpha$ and $\beta$. The plates indicate replications. For example, the document $d$ repeats $D$ times in the corpus, the word tokens $w_n$ repeats $N_d$ times in the document $d$, the vocabulary size is $W$, and there are $K$ topics. The joint probability of LDA [8] in Fig. 7.1a is

$$P(\mathbf{x}, \mathbf{z}, \boldsymbol{\theta}, \boldsymbol{\phi} | \alpha, \beta) \propto \prod_{k=1}^{K} P(\phi_k | \beta) \prod_{d=1}^{D} P(\theta_d | \alpha) \prod_{i=1}^{N_d} P(w_i | z_i^k, \phi_k) P(z_i^k | \theta_d). \quad (7.1)$$

### 7.1.1 Factor Graph for the Collapsed LDA

Based on the Drichlet-Multinomial conjugacy, integrating out the multinomial parameters $\{\theta_d, \phi_k\}$ in (7.1) yields the joint probability of the collapsed LDA [26, 27],

$$P(\mathbf{x}, \mathbf{z} | \alpha, \beta) \propto \prod_{d=1}^{D} \prod_{k=1}^{K} \frac{\Gamma(\sum_{w=1}^{W} x_{w,d} z_{w,d}^k + \alpha)}{\Gamma[\sum_{k=1}^{K} (\sum_{w=1}^{W} x_{w,d} z_{w,d}^k + \alpha)]}$$

$$\times \prod_{w=1}^{W} \prod_{k=1}^{K} \frac{\Gamma(\sum_{d=1}^{D} x_{w,d} z_{w,d}^k + \beta)}{\Gamma[\sum_{w=1}^{W} (\sum_{d=1}^{D} x_{w,d} z_{w,d}^k + \beta)]}, \quad (7.2)$$

where $x_{w,d} z_{w,d}^k = \sum_{i=1}^{x_{w,d}} z_{w,d,i}^k$ can recover the topic configuration over the word tokens in Fig. 7.1a.

Comparing (7.2) with (7.1), we find that the independence assumption of the collapsed LDA becomes different from the original LDA. For example, after integrating out the multinomial parameters $\{\theta_d, \phi_k\}$, Eq. (7.2) assumes that the hidden variables $\mathbf{z}$ are *locally dependent*, which can be encoded by the factor functions as

$$f_{\theta_d}(\mathbf{x}_{.,d}, \mathbf{z}_{.,d}, \alpha) = \prod_{k=1}^{K} \frac{\Gamma(\sum_{w=1}^{W} x_{w,d} z_{w,d}^k + \alpha)}{\Gamma[\sum_{k=1}^{K} (\sum_{w=1}^{W} x_{w,d} z_{w,d}^k + \alpha)]},$$

$$f_{\phi_w}(\mathbf{x}_{w,.}, \mathbf{z}_{w,.}, \beta) = \prod_{k=1}^{K} \frac{\Gamma(\sum_{d=1}^{D} x_{w,d} z_{w,d}^k + \beta)}{\Gamma[\sum_{w=1}^{W} (\sum_{d=1}^{D} x_{w,d} z_{w,d}^k + \beta)]},$$

where $\mathbf{z}_{.,d} = \{z_{w,d}, \mathbf{z}_{-w,d}\}$ and $\mathbf{z}_{w,.} = \{z_{w,d}, \mathbf{z}_{w,-d}\}$ denote the subsets of locally dependent variables in the factor graph of Fig. 7.1b. The *neighborhood system* of the topic label $z_{w,d}^k$ is defined as $\mathbf{z}_{-w,d}^k$ and $\mathbf{z}_{w,-d}^k$, where $\mathbf{z}_{-w,d}^k$ denotes a set of topic labels associated with all word indices in the document $d$ except $w$, and $\mathbf{z}_{w,-d}^k$ denotes a set of topic labels associated with the word indices $w$ in all documents except $d$. Therefore, the joint probability (7.2) of the collapsed LDA can be rewritten as the product of factor functions [6, Eq. (8.59)] in Fig. 7.1b,

$$P(\mathbf{x}, \mathbf{z} | \alpha, \beta) \propto \prod_{d=1}^{D} f_{\theta_d}(\mathbf{x}_{\cdot,d}, \mathbf{z}_{\cdot,d}, \alpha) \prod_{w=1}^{W} f_{\phi_w}(\mathbf{x}_{w,\cdot}, \mathbf{z}_{w,\cdot}, \beta).$$

By contrast, the noncollapsed LDA in Fig. 7.1a assumes that all hidden variables $z_{w,d,i}^k$ are *conditionally independent* given variables $\{\theta_d, \phi_k\}$.

The collapsed LDA transforms the directed graph in Fig. 7.1a into a two-layer undirected factor graph in Fig. 7.1b, of which a representative fragment is shown in Fig. 7.1c. In the collapsed space $\{\mathbf{z}, \alpha, \beta\}$, the factors $\theta_d$ and $\phi_w$ are denoted by squares, and their connected variables $z_{w,d}$ are denoted by circles. The factor $\theta_d$ connects the neighboring topic labels $\{z_{w,d}^k, \mathbf{z}_{-w,d}^k\}$ at different word indices within the same document $d$, while the factor $\phi_w$ connects the neighboring topic labels $\{z_{w,d}^k, \mathbf{z}_{w,-d}^k\}$ at the same word index $w$ but in different documents. The observed word $w$ is absorbed as the index of $\phi_w$, which is similar to absorbing the observed document $d$ as the index of $\theta_d$. The denominator of $f_{\phi_w}$ is a normalization factor in terms of $k$, so we do not explicitly connect $\phi_w$ with all hidden variables for a better illustration. Readers may refer to the hypergraph representation [5] of the collapsed LDA [73] that explicitly uses a factor node or hyperedge for the denominator. Because the factors can be parameterized functions [6], both $\theta_d$ and $\phi_w$ can encode the similar information as multinomial parameters in Fig. 7.1a.

Although Fig. 7.1b captures a different independence assumption among the collapsed hidden variables $\{\mathbf{z}, \alpha, \beta\}$ from that among $\{\mathbf{z}, \theta, \phi, \alpha, \beta\}$ in Fig. 7.1a, it is still a reasonable approximation by capturing the same joint distribution (7.2) of the collapsed LDA. In this sense, we can interpret probabilistic topic modeling as a labeling problem from the factor graph perspective. Transforming the three-layer LDA in Fig. 7.1a to the approximate graphical models or variational distributions have been previously used in VB techniques [8], which minimize the Kullback–Leibler (KL) divergence between variational and true distributions to achieve the approximation.

More generally, to approximate factor graphs to other LDA-based topic models, we need to build explicit relations between their objective joint distributions in the collapsed hidden variable space. For example, we design the proper factor functions and neighborhood systems to constrain a factor graph to represent the collapsed LDA. While the factor graph cannot capture the complete independence assumptions of the noncollapsed LDA-based topic models, it is a reasonable approximation and inspires different inference methods such as BP in the next subsection. How to approximate the two-layer factor graph to multilayer HBMs in the collapsed space still remains to be studied.

## 7.1.2 Loopy Belief Propagation (BP)

The loopy BP [6] algorithms provide exact solutions for inference problems in tree-structured factor graphs but approximate solutions in factor graphs with loops in Fig. 7.1b. Rather than directly computing the conditional joint probability $p(\mathbf{z}|\mathbf{x})$, we compute the conditional marginal or posterior probability, $p(z_{w,d}^k = 1, x_{w,d}|\mathbf{z}_{-w,-d}^k, \mathbf{x}_{-w,-d})$, called *message* $\mu_{w,d}(k)$, which can be normalized using a local computation, i.e., $\sum_{k=1}^{K} \mu_{w,d}(k) = 1, 0 \le \mu_{w,d}(k) \le 1$. According to Fig. 7.1c, we obtain

$$p(z_{w,d}^k, x_{w,d}|\mathbf{z}_{-w,-d}^k, \mathbf{x}_{-w,-d}) \propto$$
$$p(z_{w,d}^k, x_{w,d}|\mathbf{z}_{-w,d}^k, \mathbf{x}_{-w,d}) p(z_{w,d}^k, x_{w,d}|\mathbf{z}_{w,-d}^k, \mathbf{x}_{w,-d}), \tag{7.3}$$

where $-w$ and $-d$ denote all word indices except $w$ and all document indices except $d$, and the notations $\mathbf{z}_{-w,d}^k$ and $\mathbf{z}_{w,-d}^k$ represent all possible neighboring topic configurations. From the message-passing view, $p(z_{w,d}^k, x_{w,d}|\mathbf{z}_{-w,d}^k, \mathbf{x}_{-w,d})$ is the neighboring message $\mu_{\theta_d \to z_{w,d}^k}(k)$ sent from the factor node $\theta_d$, and $p(z_{w,d}^k, x_{w,d}|\mathbf{z}_{w,-d}^k, \mathbf{x}_{w,-d})$ is the other neighboring message $\mu_{\phi_w \to z_{w,d}^k}(k)$ sent from the factor node $\phi_w$. Using the Bayes' rule and the joint probability (7.2), we can expand Eq. (7.3) as

$$\mu_{w,d}(k) \propto \frac{p(\mathbf{z}_{\cdot,d}^k, \mathbf{x}_{\cdot,d})}{p(\mathbf{z}_{-w,d}^k, \mathbf{x}_{-w,d})} \times \frac{p(\mathbf{z}_{w,\cdot}^k, \mathbf{x}_{w,\cdot})}{p(\mathbf{z}_{w,-d}^k, \mathbf{x}_{w,-d})}$$

$$\propto \frac{\sum_{-w} x_{-w,d} z_{-w,d}^k + \alpha}{\sum_{k=1}^{K}(\sum_{-w} x_{-w,d} z_{-w,d}^k + \alpha)}$$

$$\times \frac{\sum_{-d} x_{w,-d} z_{w,-d}^k + \beta}{\sum_{w=1}^{W}(\sum_{-d} x_{w,-d} z_{w,-d}^k + \beta)}, \tag{7.4}$$

where the property, $\Gamma(x + 1) = x\Gamma(x)$, is used to cancel the common terms in both nominator and denominator [27].[1] We find that Eq. (7.4) updates the message on the variable $z_{w,d}^k$ if its neighboring topic configuration $\{\mathbf{z}_{-w,d}^k, \mathbf{z}_{w,-d}^k\}$ is known. However, due to uncertainty, we know only the neighboring messages rather than the precise topic configuration. So, we replace topic configurations by corresponding messages in Eq. (7.4) and obtain the following message update equation,

$$\mu_{w,d}(k) \propto \frac{\mu_{-w,d}(k) + \alpha}{\sum_k[\mu_{-w,d}(k) + \alpha]} \times \frac{\mu_{w,-d}(k) + \beta}{\sum_w[\mu_{w,-d}(k) + \beta]}, \tag{7.5}$$

where

---

[1] Because $\mu_{w,d}(k) \approx \mu_{w,d,i}(k)$, the difference between numerator and denominator in Gamma functions is one. Canceling the common terms yields (7.4) by subtracting a relatively small number $x_{w,d} z_{w,d}^k$ in both numerator and denominator.

> **input**     : $\mathbf{x}, K, T, \alpha, \beta$.
> **output**   : $\theta_d, \phi_w$.
> $\mu_{w,d}^1(k) \leftarrow$ initialization and normalization;
> **for** $t \leftarrow 1$ **to** $T$ **do**
> $\quad \mu_{w,d}^{t+1}(k) \propto \dfrac{\mu_{-w,d}^t(k)+\alpha}{\sum_k [\mu_{-w,d}^t(k)+\alpha]} \times \dfrac{\mu_{w,-d}^t(k)+\beta}{\sum_w [\mu_{w,-d}^t(k)+\beta]}$;
> **end**
> $\theta_d(k) \leftarrow [\boldsymbol{\mu}_{\cdot,d}(k)+\alpha]/\sum_k [\boldsymbol{\mu}_{\cdot,d}(k)+\alpha]$;
> $\phi_w(k) \leftarrow [\boldsymbol{\mu}_{w,\cdot}(k)+\beta]/\sum_w [\boldsymbol{\mu}_{w,\cdot}(k)+\beta]$;

**Fig. 7.2** The synchronous BP for LDA

$$\boldsymbol{\mu}_{-w,d}(k) = \sum_{-w} x_{-w,d}\mu_{-w,d}(k), \tag{7.6}$$

$$\boldsymbol{\mu}_{w,-d}(k) = \sum_{-d} x_{w,-d}\mu_{w,-d}(k). \tag{7.7}$$

Messages are passed from variables to factors, and in turn from factors to variables until convergence or the maximum number of iterations is reached. Notice that we need only pass messages for $x_{w,d} \neq 0$. Because $\mathbf{x}$ is a very sparse matrix, the message update equation (7.5) is computationally fast by sweeping all nonzero elements in the sparse matrix $\mathbf{x}$.

Based on the normalized messages, we can estimate the multinomial parameters $\theta$ and $\phi$ by the expectation–maximization (EM) algorithm [27]. Using the Dirichlet-Multinomial conjugacy and Bayes' rule, we express the marginal Dirichlet distributions on parameters,

$$p(\theta_d) = \mathrm{Dir}(\theta_d | \boldsymbol{\mu}_{\cdot,d}(k) + \alpha), \tag{7.8}$$

$$p(\phi_w) = \mathrm{Dir}(\phi_w | \boldsymbol{\mu}_{w,\cdot}(k) + \beta). \tag{7.9}$$

The M-step maximizes (7.8) and (7.9) with respect to $\theta_d$ and $\phi_w$, resulting in the following point estimates of multinomial parameters,

$$\theta_d(k) = \frac{\boldsymbol{\mu}_{\cdot,d}(k) + \alpha}{\sum_k [\boldsymbol{\mu}_{\cdot,d}(k) + \alpha]}, \tag{7.10}$$

$$\phi_w(k) = \frac{\boldsymbol{\mu}_{w,\cdot}(k) + \beta}{\sum_w [\boldsymbol{\mu}_{w,\cdot}(k) + \beta]}. \tag{7.11}$$

To estimate hyperparameters based on inferred messages, we refer readers to [2, 27].

To implement the BP algorithm, we must choose either the synchronous or the asynchronous update schedule to pass messages [57]. Figure 7.2 shows the synchronous message update schedule. At each iteration $t$, each variable uses the incoming messages in the previous iteration $t - 1$ to calculate the current message. Once every variable computes its message, the message is passed to the neighboring variables and used to compute messages in the next iteration $t + 1$. An alternative is the asynchronous message update schedule. It updates the message of each variable immediately. The updated message is immediately used to compute other neighboring messages at each iteration $t$. The asynchronous update schedule often passes messages faster across variables, which causes the BP algorithm converge faster than the synchronous update schedule. Another termination condition for convergence is that the change of the multinomial parameters [8] is less than a predefined threshold $\lambda$, for example, $\lambda = 0.00001$ [28].

Figure 7.3a shows the hypergraph for the joint probability (7.2). There are three types of hyperedges $\{\theta_d, \phi_w, \gamma_k\}$ denoted by the yellow, green and red rectangles, respectively. For each column of $\mathbf{z}_{W \times D}$, the hyperedge $\theta_d$, which corresponds to the first term in (7.2), connects the variable $z_{w,d}^k$ with the subset of variables $\mathbf{z}_{-w,d}^k$ and the hyperparameter $\alpha$ within the document $d$. For each row of $\mathbf{z}_{W \times D}$, the hyperedge $\phi_w$, which corresponds to the second term in (7.2), connects the variable $z_{w,d}^k$ with the subset of variables $\mathbf{z}_{w,-d}^k$ and the hyperparameter $\beta$ at the same word $w$ in the vocabulary. Finally, the hyperedge $\gamma_k$, which corresponds to the third term in (7.2), connects the variables $z_{w,d}^k$ with all other variables $\mathbf{z}_{-w,-d}^k$ and the hyperparameter $\beta$ on the same topic $k$. The notations $-w$ and $-d$ denote all word indices except $w$ and all document indices except $d$, and the notations $\mathbf{z}_{-w,d}^k$, $\mathbf{z}_{w,-d}^k$ and $\mathbf{z}_{-w,-d}^k$ represent all neighboring labeling configurations through three types of hyperedges $\{\theta_d, \phi_w, \gamma_k\}$, respectively. Therefore, the hypergraph can completely describe the local dependencies of the topic configuration $\mathbf{z}$ in (7.2).

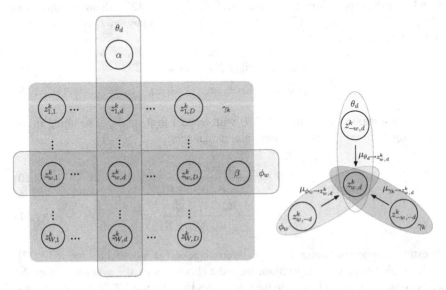

**Fig. 7.3** **a** Hypergraph representation for the collapsed LDA and **b** message passing on hypergraph

### 7.1.3 An Alternative View of BP

We may also adopt one of the BP instantiations, the sum–product algorithm [6], to infer $\mu_{w,d}(k)$. For convenience, we will not include the observation $x_{w,d}$ in the formulation. Figure 7.1c shows the message passing from two factors $\theta_d$ and $\phi_w$ to the variable $z_{w,d}^k$, where the arrows denote the message-passing directions. Based on the smoothness prior, we encourage only $K$ smooth topic configurations without considering all other possible configurations. The message $\mu_{w,d}(k)$ is proportional to the product of both incoming messages from factors,

$$\mu_{w,d}(k) \propto \mu_{\theta_d \to z_{w,d}^k}(k) \times \mu_{\phi_w \to z_{w,d}^k}(k). \tag{7.12}$$

Equation (7.12) has the same meaning with (7.3). The messages from factors to variables are the sum of all incoming messages from the neighboring variables [6],

$$\mu_{\theta_d \to z_{w,d}^k}(k) = f_{\theta_d} \prod_{-w} \mu_{-w,d}(k)\alpha, \tag{7.13}$$

$$\mu_{\phi_w \to z_{w,d}^k}(k) = f_{\phi_w} \prod_{-d} \mu_{w,-d}(k)\beta, \tag{7.14}$$

where $\alpha$ and $\beta$ can be viewed as the pseudo-messages, and $f_{\theta_d}$ and $f_{\phi_w}$ are the factor functions that encourage or penalize the incoming messages. Note that the sum operation in the sum-produce algorithm is not needed because LDA encourages only $K$ smooth topic configurations.

In practice, however, Eqs. (7.13) and (7.14) often cause the product of multiple incoming messages close to zero. To avoid arithmetic underflow, we approximate the product operation by the sum operation of incoming messages, because when the product value increases the sum value also increases,

$$\prod_{-w} \mu_{-w,d}(k)\alpha \propto \sum_{-w} \mu_{-w,d}(k) + \alpha, \tag{7.15}$$

$$\prod_{-d} \mu_{w,-d}(k)\beta \propto \sum_{-d} \mu_{w,-d}(k) + \beta. \tag{7.16}$$

Such approximations transform the sum–product to the sum-sum algorithm, similar to the relaxation labeling algorithm with good performance.

The normalized message $\mu_{w,d}(k)$ is multiplied by the number of word counts $x_{w,d}$ during the propagation, i.e., $x_{w,d}\mu_{w,d}(k)$. In this sense, $x_{w,d}$ can be viewed as the weight of $\mu_{w,d}(k)$ during the propagation, where the bigger $x_{w,d}$ corresponds to the larger influence of its message to those of its neighbors. Thus, the topics may be distorted by those documents with greater word counts. To avoid this problem, we may choose another weighting scheme like term frequency (TF) or term frequency-inverse document frequency (TF-IDF) for *weighted* belief propagation. Therefore, BP can not only handle *discrete* data, but also process *continuous* data like TF-IDF. The factor graph in Fig. 7.1b can be extended to describe both discrete and continuous data in general, while LDA in Fig. 7.1a focuses only on generating discrete data. When the factor graph processes the discrete data, it reduces to the collapsed LDA as a special case.

We can design the factor functions arbitrarily to encourage or penalize local topic labeling configurations based on our prior knowledge. From Fig. 7.1a, LDA solves the topic modeling problem according to three intrinsic assumptions:

1. Co-occurrence: Different word indices within the same document tend to be associated with the same topic labels.
2. Smoothness: The same word indices in different documents are likely to be associated with the same topic labels.
3. Clustering: All word indices do not tend to associate with the same topic labels.

The first assumption is determined by the document-specific topic proportion $\theta_d(k)$, where it is more likely to assign a topic label $z^k_{w,d} = 1$ to the word index $w$ if the topic $k$ is more frequently assigned to other word indices in the document $d$. Similarly, the second assumption is based on the topic-specific multinomial distribution $\phi_k(w)$, which implies a higher likelihood to associate the word index $w$ with the topic label $z^k_{w,d} = 1$ if $k$ is more frequently assigned to the same word index $w$ in other documents except $d$. The third assumption avoids grouping all word indices into one topic through normalizing $\phi_k(w)$ in terms of all word indices. If most word indices are associated with the topic $k$, the multinomial parameter $\phi_k$ will become too small to allocate the topic $k$ to these word indices.

According to the above assumptions, we design $f_{\theta_d}$ and $f_{\phi_w}$ over messages as

$$f_{\theta_d}(\boldsymbol{\mu}_{\cdot,d}, \alpha) = \frac{1}{\sum_k [\boldsymbol{\mu}_{-w,d}(k) + \alpha]}, \qquad (7.17)$$

$$f_{\phi_w}(\boldsymbol{\mu}_{w,\cdot}, \beta) = \frac{1}{\sum_w [\boldsymbol{\mu}_{w,-d}(k) + \beta]}. \qquad (7.18)$$

Equation (7.17) normalizes the incoming messages by the total number of messages for all topics associated with the document $d$ to make outgoing messages comparable across documents. Equation (7.18) normalizes the incoming messages by the total number of messages for all words in the vocabulary to make outgoing messages comparable across vocabulary words. Notice that (7.13) and (7.14) realize the first two assumptions, and (7.18) encodes the third assumption of topic modeling. The similar normalization technique to avoid partitioning all data points into one cluster has been used in the classic normalized cuts algorithm for image segmentation [54]. Combining (7.12)–(7.18) will yield the same message update Eq. (7.5). To estimate parameters $\theta_d$ and $\phi_w$, we use the joint marginal distributions (7.13) and (7.14) of the set of variables belonging to factors $\theta_d$ and $\phi_w$ including the variable $z^k_{w,d}$, which produce the same point estimation equations (7.10) and (7.11).

### 7.1.4 Simplified BP (siBP)

We may simplify the message update equation (7.5). Substituting Eqs. (7.10) and (7.11) into (7.5) yields the approximate message update equation,

$$\mu_{w,d}(k) \propto \theta_d(k) \times \phi_w(k), \tag{7.19}$$

which includes the current message $\mu_{w,d}(k)$ in both numerator and denominator in (7.5). In many real-world topic modeling tasks, a document often contains many different word indices, and the same word index appears in many different documents. So, at each iteration, Eq. (7.19) deviates slightly from (7.5) after adding the current message to both numerator and denominator. Such slight difference may be enlarged after many iterations in Fig. 7.2 due to accumulation effects, leading to different estimated parameters.

Intuitively, Eq. (7.19) implies that if the topic $k$ has a higher proportion in the document $d$, and it has the a higher likelihood to generate the word index $w$, it is more likely to allocate the topic $k$ to the observed word $x_{w,d}$. This allocation scheme in principle follows the three intrinsic topic modeling assumptions in the Sect. 7.1.3. Figure 7.4 shows the MATLAB code for the simplified BP (siBP).

### 7.1.5 Relationship to Previous Algorithms

Here we discuss some intrinsic relations between BP with three state-of-the-art LDA learning algorithms such as VB [8], GS [26], and zero-order approximation CVB (CVB0) [2, 3] within the unified message-passing framework. The message update scheme is an instantiation of the E-step of EM algorithm [18], which has been widely used to infer the marginal probabilities of hidden variables in various graphical models according to the maximum-likelihood estimation [6] (e.g., the E-step inference for GMMs [84], the forward–backward algorithm for HMMs, and the probabilistic relaxation labeling algorithm for MRF [85]). After the E-step, we estimate the optimal parameters using the updated messages and observations at the M-step of EM algorithm.

VB is a variational message-passing method [65] that uses a set of factorized variational distributions $q(\mathbf{z})$ to approximate the joint distribution (7.2) by minimizing the KL divergence between them. Employing the Jensen's inequality makes the approximate variational distribution an adjustable lower bound on the joint distribution, so that maximizing the joint probability is equivalent to maximizing the lower bound by tuning a set of variational parameters. Because there is always a gap between the lower bound and the true joint distribution, VB introduces bias when learning LDA. The variational message update equation is

$$\mu_{w,d}(k) \propto \frac{\exp[\Psi(\mu_{\cdot,d}(k) + \alpha)]}{\exp[\Psi(\sum_k [\mu_{\cdot,d}(k) + \alpha])]} \times \frac{\mu_{w,\cdot}(k) + \beta}{\sum_w [\mu_{w,\cdot}(k) + \beta]}, \tag{7.20}$$

```
function [phi, theta] = siBP(X, K, T, ALPHA, BETA)

% X is a W*D sparse matrix.
% W is the vocabulary size.
% D is the number of documents.
% The element of X is the word count 'xi'.
% 'wi' and 'di' are word and document indices.
% K is the number of topics.
% T is the number of iterations.
% mu is a matrix with K rows for topic messages.
% phi is a K*W matrix.
% theta is a K*D matrix.
% ALPHA and BETA are hyperparameters.
%
% normalize(A,dim) returns the normalized values
% (sum to one) of the elements along different
% dimensions of an array.

[wi,di,xi] = find(X);
% random initialization
mu = normalize(rand(K,nnz(X)),1);
% simplified belief propagation
for t = 1:T
    for k = 1:K
        md(k,:) = accumarray(di,xi'.*mu(k,:));
        mw(k,:) = accumarray(wi,xi'.*mu(k,:));
    end
    theta = normalize(md+ALPHA,1); %Eq.(9)
    phi = normalize(mw+BETA,2); %Eq.(10)
    mu = normalize(theta(:,di).*phi(:,wi),1); %Eq.(18)
end

return
```

**Fig. 7.4** The MATLAB code for siBP

which resembles the synchronous BP (7.5) but with two major differences. First, VB uses complicated digamma functions $\Psi(\cdot)$, which not only introduces bias [2] but also slows down the message updating. Second, VB uses a different variational EM schedule. At the E-step, it simultaneously updates both variational messages and parameter of $\theta_d$ until convergence, holding the variational parameter of $\phi$ fixed. At the M-step, VB updates only the variational parameter of $\phi$.

The message update equation of GS is

$$\mu_{w,d,i}(k) \propto \frac{n_{\cdot,d}^{-i}(k) + \alpha}{\sum_k [n_{\cdot,d}^{-i}(k) + \alpha]} \times \frac{n_{w,\cdot}^{-i}(k) + \beta}{\sum_w [n_{w,\cdot}^{-i}(k) + \beta]}, \qquad (7.21)$$

where $n_{\cdot,d}^{-i}(k)$ is the total number of topic labels $k$ in the document $d$ except the topic label on the current word token $i$, and $n_{w,\cdot}^{-i}(k)$ is the total number of topic labels $k$ of the word $w$ except the topic label on the current word token $i$. Equation (7.21) resembles the asynchronous BP implementation (7.5) but with two subtle differences. First, GS randomly samples the current topic label $z_{w,d,i}^{k} = 1$ from the message $\mu_{w,d,i}(k)$, which truncates all $K$-tuple message values to zeros except the sampled topic label $k$. Such information loss introduces bias when learning LDA. Second, GS must sample a topic label for each word token, which repeats $x_{w,d}$ times for the word index $\{w, d\}$. The sweep of the entire word tokens rather than word index restricts GS's scalability to large-scale document repositories containing billions of word tokens.

CVB0 is exactly equivalent to our asynchronous BP implementation but based on word tokens. Previous empirical comparisons [2] advocated the CVB0 algorithm for LDA within the approximate mean-field framework [3] closely connected with the proposed BP. Here, we clearly explain that the superior performance of CVB0 has been largely attributed to its asynchronous BP implementation. Our experiments also support that the message passing over word indices instead of tokens will produce comparable or even better topic modeling performance but with significantly smaller computational costs.

Equation (7.19) also reveals that siBP is a probabilistic matrix factorization algorithm that factorizes the document-word matrix, $\mathbf{x} = [x_{w,d}]_{W \times D}$, into a matrix of document-specific topic proportions, $\boldsymbol{\theta} = [\theta_d(k)]_{K \times D}$, and a matrix of vocabulary word-specific topic proportions, $\boldsymbol{\phi} = [\phi_w(k)]_{K \times W}$, i.e., $\mathbf{x} \sim \boldsymbol{\phi}^T \boldsymbol{\theta}$. We see that the larger number of word counts $x_{w,d}$ corresponds to the higher likelihood $\sum_k \theta_d(k)\phi_w(k)$. From this point of view, the multinomial principle component analysis (PCA) [12] describes some intrinsic relations among LDA, PLSA [29], and non-negative matrix factorization (NMF) [34]. Equation (7.19) is the same as the E-step update for PLSA except that the parameters $\theta$ and $\phi$ are smoothed by the hyperparameters $\alpha$ and $\beta$ to prevent overfitting.

VB, BP, and siBP have the same time complexity $\mathcal{O}(KDW_dT)$, but GS and CVB0 have $\mathcal{O}(KDN_dT)$, where $W_d$ is the average vocabulary size, $N_d$ is the average number of word tokens per document, and $T$ is the number of learning iterations. Because VB and BP pass the entire $K$-tuple messages, their space complexity for message passing increases linearly with the number of topics $K$. When $K$ is very large, VB and BP often require a significantly higher memory usage than GS.

## 7.1.6 Belief Propagation for ATM

Author-topic models (ATM) [51] depict each author of the document as a mixture of probabilistic topics, and have found important applications in matching chapters with reviewers [74]. Figure 7.5a shows the generative graphical representation for ATM, which first uses a document-specific uniform distribution $u_d$ to generate an

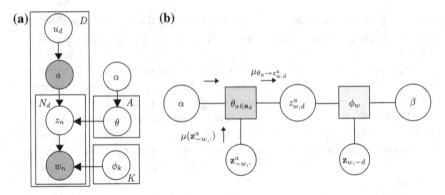

**Fig. 7.5  a** The three-layer graphical representation [51] and **b** two-layer factor graph of ATM

author index $a$, $1 \leq a \leq A$, and then uses the author-specific topic proportions $\theta_a$ to generate a topic label $z_{w,d}^k = 1$ for the word index $w$ in the document $d$. The plate on $\theta$ indicates that there are $A$ unique authors in the corpus. The document often has multiple coauthors. ATM randomly assigns one of the observed author indices to each word in the document based on the document-specific uniform distribution $u_d$. However, it is more reasonable that each word $x_{w,d}$ is associated with an author index $a \in \mathbf{a}_d$ from the multinomial rather than uniform distribution, where $\mathbf{a}_d$ is a set of author indices of the document $d$. As a result, each topic label takes two variables $z_{w,d}^{a,k} = \{0, 1\}$, $\sum_{a,k} z_{w,d}^{a,k} = 1$, $a \in \mathbf{a}_d$, $1 \leq k \leq K$, where $a$ is the author index and $k$ is the topic index attached to the word.

We transform Fig. 7.5a to the factor graph representation of ATM in Fig. 7.5b. As with Fig. 7.1c, we absorb the observed author index $a \in \mathbf{a}_d$ of the document $d$ as the index of the factor $\theta_{a \in \mathbf{a}_d}$. The notation $\mathbf{z}_{-w,\cdot}^a$ denotes all labels connected with the authors $a \in \mathbf{a}_d$ except those for the word index $w$. The only difference between ATM and LDA is that the author $a \in \mathbf{a}_d$, instead of the document $d$, connects the labels $z_{w,d}^a$ and $\mathbf{z}_{-w,\cdot}^a$. As a result, ATM encourages topic smoothness among labels $z_{w,d}^a$ attached to the same author $a$ instead of the same document $d$.

### 7.1.6.1  Inference and Parameter Estimation

Unlike passing the $K$-tuple message $\mu_{w,d}(k)$ in Fig. 7.2, the BP algorithm for learning ATM passes the $|\mathbf{a}_d| \times K$-tuple message vectors $\mu_{w,d}^a(k)$, $a \in \mathbf{a}_d$ through the factor $\theta_{a \in \mathbf{a}_d}$ in Fig. 7.5b, where $|\mathbf{a}_d|$ is the number of authors in the document $d$. Nevertheless, we can still obtain the $K$-tuple word topic message $\mu_{w,d}(k)$ by marginalizing the message $\mu_{w,d}^a(k)$ in terms of the author variable $a \in \mathbf{a}_d$ as follows,

$$\mu_{w,d}(k) = \sum_{a \in \mathbf{a}_d} \mu_{w,d}^a(k). \tag{7.22}$$

Since Figs. 7.1c, 7.5b have the same right half part, the message-passing equation from the factor $\phi_w$ to the variable $z_{w,d}$ and the parameter estimation equation for $\phi_w$ in Fig. 7.5b remain the same as (7.5) and (7.11) based on the marginalized word topic message in (7.22). Thus, we only need to derive the message-passing equation from the factor $\theta_{a \in \mathbf{a}_d}$ to the variable $z_{w,d}^a$ in Fig. 7.5b. Because of the topic smoothness prior, we design the factor function as follows,

$$f_{\theta_a} = \frac{1}{\sum_k [\mu_{-w,-d}^a(k) + \alpha]}, \qquad (7.23)$$

where $\mu_{-w,-d}^a(k) = \sum_{-w,-d} x_{w,d}^a \mu_{w,d}^a(k)$ denotes the sum of all incoming messages attached to the author index $a$ and the topic index $k$ excluding $x_{w,d}^a \mu_{w,d}^a(k)$. Likewise, Eq. (7.23) normalizes the incoming messages attached to the author index $a$ in terms of the topic index $k$ to make outgoing messages comparable for different authors $a \in \mathbf{a}_d$. Similar to (7.13), we derive the message passing $\mu_{f_{\theta_a} \to z_{w,d}^a}$ through adding all incoming messages evaluated by the factor function (7.23).

Multiplying two messages from factors $\theta_{a \in \mathbf{a}_d}$ and $\phi_w$ yields the message update equation as follows,

$$\mu_{w,d}^a(k) \propto \frac{\mu_{-w,-d}^a(k) + \alpha}{\sum_k [\mu_{-w,-d}^a(k) + \alpha]} \times \frac{\mu_{w,-d}(k) + \beta}{\sum_w [\mu_{w,-d}(k) + \beta]}. \qquad (7.24)$$

Notice that the $|\mathbf{a}_d| \times K$-tuple message $\mu_{w,d}^a(k), a \in \mathbf{a}_d$ is normalized in terms of all combinations of $\{a, k\}, a \in \mathbf{a}_d, 1 \le k \le K$. Based on the normalized messages, the author-specific topic proportion $\theta_a(k)$ can be estimated from the sum of all incoming messages including $\mu_{w,d}^a$ evaluated by the factor function $f_{\theta_a}$ as follows,

$$\theta_a(k) = \frac{\mu_{\cdot,\cdot}^a(k) + \alpha}{\sum_k [\mu_{\cdot,\cdot}^a(k) + \alpha]}. \qquad (7.25)$$

As a summary, Fig. 7.6 shows the synchronous BP algorithm for learning ATM. The difference between Fig. 7.2 and Fig. 7.6 is that Fig. 7.2 considers the author index $a$ as the label for each word. At each iteration, the computational complexity is $\mathcal{O}(KDW_d A_d T)$, where $A_d$ is the average number of authors per document.

## 7.1.7 Belief Propagation for RTM

Network data, such as citation and coauthor networks of documents [74, 75], tag networks of documents and images [76], hyperlinked networks of web pages, and social networks of friends, exist pervasively in data mining and machine learning. The probabilistic relational topic modeling of network data can provide both useful predictive models and descriptive statistics [16].

In Fig. 7.7a, relational topic models (RTM) [16] represent entire document topics by the mean value of the document topic proportions, and use Hadamard product of

$$\text{input} \quad : \mathbf{x}, \mathbf{a}_d, K, T, \alpha, \beta.$$

$$\text{output} \quad : \theta_a, \phi_w.$$

$$\mu_{w,d}^{a,1}(k), a \in \mathbf{a}_d, \leftarrow \text{initialization and normalization;}$$

$$\textbf{for } t \leftarrow 1 \textbf{ to } T \textbf{ do}$$

$$\left| \quad \mu_{w,d}^{a,t+1}(k) \propto \frac{\mu_{-w,-d}^{a,t}(k)+\alpha}{\sum_k [\mu_{-w,-d}^{a,t}(k)+\alpha]} \times \frac{\mu_{w,-d}^{t}(k)+\beta}{\sum_w [\mu_{w,-d}^{t}(k)+\beta]}; \right.$$

$$\textbf{end}$$

$$\theta_a(k) \leftarrow [\boldsymbol{\mu}_{\cdot,\cdot}^{a}(k)+\alpha]/\sum_k[\boldsymbol{\mu}_{\cdot,\cdot}^{a}(k)+\alpha];$$

$$\phi_w(k) \leftarrow [\boldsymbol{\mu}_{w,\cdot}(k)+\beta]/\sum_w[\boldsymbol{\mu}_{w,\cdot}(k)+\beta];$$

**Fig. 7.6** The synchronous BP for ATM

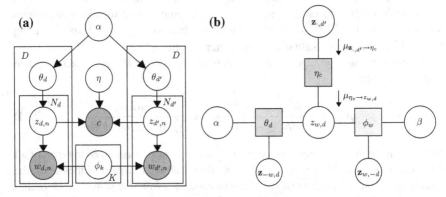

**Fig. 7.7** **a** The three-layer graphical representation [16] and **b** two-layer factor graph of RTM

mean values $\bar{z}_d \circ \bar{z}_{d'}$ from two linked documents $\{d, d'\}$ as link features, which are learned by the generalized linear model (GLM) $\eta$ to generate the observed binary citation link variable $c = 1$. Besides, all other parts in RTM remain the same as LDA.

We transform Fig. 7.7a to the factor graph Fig. 7.7b by absorbing the observed link index $c \in \mathbf{c}$, $1 \leq c \leq C$ as the index of the factor $\eta_c$. Each link index connects a document pair $\{d, d'\}$, and the factor $\eta_c$ connects word topic labels $z_{w,d}$ and $\mathbf{z}_{\cdot,d'}$ of the document pair. Besides encoding the topic smoothness, RTM explicitly describes the topic structural dependencies between the pair of linked documents $\{d, d'\}$ using the factor function $f_{\eta_c}(\cdot)$.

### 7.1.7.1 Inference and Parameter Estimation

In Fig. 7.7, the messages from the factors $\theta_d$ and $\phi_w$ to the variable $z_{w,d}$ are the same as LDA in (7.13) and (7.14). Thus, we only need to derive the message-passing equation from the factor $\eta_c$ to the variable $z_{w,d}$.

**Fig. 7.8** The synchronous
BP for RTM

input    : $\mathbf{w}, \mathbf{c}, \xi, K, T, \alpha, \beta$.
output  : $\theta_a, \phi_w$.
$\mu^1_{w,d}(k) \leftarrow$ initialization and normalization;
**for** $t \leftarrow 1$ **to** $T$ **do**
$\qquad \mu^{t+1}_{w,d}(k) \propto$
$\qquad [(1-\xi)\mu^t_{\theta_d \rightarrow z_{w,d}}(k) + \xi\mu^t_{\eta_c \rightarrow z_{w,d}}(k)] \times \mu^t_{\phi_w \rightarrow z_{w,d}}(k);$
**end**
$\theta_d(k) \leftarrow [\mu_{\cdot,d}(k) + \alpha]/\sum_k[\mu_{\cdot,d}(k) + \alpha];$
$\phi_w(k) \leftarrow [\mu_{w,\cdot}(k) + \beta]/\sum_w[\mu_{w,\cdot}(k) + \beta];$

We design the factor function $f_{\eta_c}(\cdot)$ for linked documents as follows,

$$f_{\eta_c}(k|k') = \frac{\sum_{\{d,d'\}} \boldsymbol{\mu}_{\cdot,d}(k)\boldsymbol{\mu}_{\cdot,d'}(k')}{\sum_{\{d,d'\},k'} \boldsymbol{\mu}_{\cdot,d}(k)\boldsymbol{\mu}_{\cdot,d'}(k')}, \qquad (7.26)$$

which depicts the likelihood of topic label $k$ assigned to the document $d$ when its
linked document $d'$ is associated with the topic label $k'$. Notice that the designed
factor function does not follow the GLM for link modeling in the original RTM [16]
because the GLM makes inference slightly more complicated. However, similar to
the GLM, Eq. (7.26) is also able to capture the topic interactions between two linked
documents $\{d, d'\}$ in document networks. Instead of smoothness prior encoded by
factor functions (7.17) and (7.18), it describes arbitrary topic dependencies $\{k, k'\}$
of linked documents $\{d, d'\}$.

Based on the factor function (7.26), we resort to the sum–product algorithm to
calculate the message,

$$\mu_{\eta_c \rightarrow z_{w,d}}(k) = \sum_{d'} \sum_{k'} f_{\eta_c}(k|k')\boldsymbol{\mu}_{\cdot,d'}(k'), \qquad (7.27)$$

where we use the sum rather than the product of messages from all linked documents
$d'$ to avoid arithmetic underflow. The standard sum–product algorithm requires the
product of all messages from factors to variables. However, in practice, the direct
product operation cannot balance the messages from different sources. For example,
the message $\mu_{\theta_d \rightarrow z_{w,d}}$ is from the neighboring words within the same document
$d$, while the message $\mu_{\eta_c \rightarrow z_{w,d}}$ is from all linked documents $d'$. If we pass the
product of these two types of messages, we cannot distinguish which one influences
more on the topic label $z_{w,d}$. Hence, we use the weighted sum of two types of
messages,

$$\mu(z_{w,d} = k) \propto [(1 - \xi)\mu_{\theta_d \rightarrow z_{w,d}}(k) + \xi\mu_{\eta_c \rightarrow z_{w,d}}(k)] \times \mu_{\phi_w \rightarrow z_{w,d}}(k), \qquad (7.28)$$

where $\xi \in [0, 1]$ is the weight to balance two messages $\mu_{\theta_d \rightarrow z_{w,d}}$ and $\mu_{\eta_c \rightarrow z_{w,d}}$. When
there are no link information $\xi = 0$, Eq. (7.28) reduces to (7.5) so that RTM reduces
to LDA. Figure 7.8 shows the synchronous BP algorithm for learning RTM. Given

the inferred messages, the parameter estimation equations remain the same as (7.10) and (7.11). The computational complexity at each iteration is $\mathcal{O}(K^2 C D W_d T)$, where $C$ is the total number of links in the document network.

## 7.2 Speedup Topic Modeling

The probabilistic topic modeling techniques divide the nonzero elements in document-word matrix into several thematic groups called topics, which have have been applied widely in text mining, computer vision, and computational biology [9]. As the simplest topic model, latent Dirichlet allocation (LDA) [8] represents each document as a mixture of topics and each topic as a multinomial distribution over a fixed vocabulary. From the labeling point of view, LDA assigns the hidden topic labels to explain the observed words in document-word matrix [76]. This labeling process defines a joint probability distribution over the hidden labels and the observed words. Employing the Bayes' rule, we can infer topic labels from observed words by computing the posterior distribution of the hidden variables given the observed variables from their joint probability. Such inference techniques have been widely used for learning probabilistic graphical models within the Bayesian framework [6]. Recent approximate inference methods for LDA [2] fall broadly into three categories: variational Bayes (VB) [8], collapsed Gibbs sampling (GS) [26], and loopy belief propagation (BP) [76].

However, all above batch LDA inference algorithms require multiple iterations of scanning the entire corpus and searching the complete topic space. So, the computational cost increases linearly with the number of documents $D$, the number of topics $K$, and the number of training iterations $T$. For massive corpora containing the large number of topics, these batch LDA algorithms are often inefficient for fast topic modeling because of the high per-iteration cost. For example, when $D = 10^6$, $K = 10^3$, and $T = 500$, GS consumes around forty hours to infer the hidden topics. Therefore, we need faster batch LDA algorithms for real-world massive corpora.

In this section, we introduce a fast batch LDA algorithm, active belief propagation (ABP). The basic idea is to actively select a subset of corpus and search a subset of topic space at each training iteration. ABP is based on the residual BP (RBP) [21] framework that uses an informed scheduling for asynchronous message passing. First, we define the message residual as the absolute difference between messages at successive training iterations, and the residuals are used to evaluate the convergence speed of messages. The larger the residuals the faster the convergence speed. Second, through sorting residuals at each training iteration, we actively select those documents and topics that contribute the largest residuals for message updating and passing. For example, if ABP selects 10 % documents and 10 % topics for message passing at each iteration, it will reach as high as 100 times faster than BP. Extensive experiments on four real-world corpora have demonstrated that ABP can achieve a significant speedup with a comparable accuracy as that achieved by current state-of-the-art batch LDA algorithms.

The proposed ABP algorithm can be interpreted as a fast expectation–maximization (EM) algorithm [18]. At the E-step, ABP infers and updates the subset of messages with the largest residuals. At the M-step, ABP maximizes the likelihood by estimating parameters based on the subset of updated messages. After multiple iterations, it often converges at a local maximum of the joint probability. Because the residuals reflect the convergence speed of messages, unlike lazy EM [59], ABP focuses all computational resources on inferring and updating those fast-converging messages at each iteration. Indeed, the document message residuals approximately follow Zipf's law [90], where 20 % documents contribute almost 80 % residuals. This phenomenon informs us that scanning only a small subset of documents at each iteration by ABP can ensure almost the same topic modeling accuracy as BP. Since BP can be applied to training other LDA-based topic models [76], we believe that the basic idea of ABP can be also extended to speed up more LDA-based topic modeling algorithms.

## 7.2.1 Fast Topic Modeling Techniques

Current fast batch LDA algorithms can handle the large number of topics $K$. For example, fast GS (FGS) [48] introduces the upper bound on the normalization factor using the Hölder's inequality. Without visiting all possible $K$ topics, FGS is able to draw the target topic sample from the approximate posterior probability normalized by the upper bound. By refining the upper bound to approximate the true normalization factor, FGS samples the topic label equal to that drawn from the true posterior probability as GS does. In this sense, FGS yields exactly the same result as GS but with much less computation. Due to the sparseness of the $K$-tuple messages, sparse GS (SGS) [70] partitions the $K$-tuple messages into three parts from which a random topic label can be efficiently sampled. As a result, SGS avoids additional computations in GS during the sampling process.

Online LDA algorithms [1, 4, 13, 28, 42, 61, 70] partition the entire corpus into a set of mini-batches. They infer topic distributions for each unseen mini-batch based on previous topic distributions, and thus converge significantly faster than batch algorithms, saving enormous training time by reducing the total number of iterations $T$. In addition, online LDA algorithms discard each mini-batch after one look, and thus require a constant memory usage. For example, incremental LDA (ILDA) [13] periodically resamples the topic assignments for previously analyzed words, or uses particle filtering instead of GS to assign the current topic based on previous topic configurations. Despite faster speed, ILDA does not perform as well as GS. Online GS (OGS) [70] extends SGS to infer topics from unseen data streams. Online VB (OVB) [28] combines VB with the online stochastic optimization framework [11], and can converge to a local optimum of the VB objective function. However, OVB requires manually tuning several parameters including the mini-batch size that determines the best topic modeling performance based on heuristics. Residual VB (RVB) [61] dynamically schedules mini-batches to speed up convergence of OVB. Sampled online inference (SOI) combines SGS with the scalability of online stochastic inference [42]. However, both RVB and SOI cannot process data streams because they reuse previously seen mini-batches.

Parallel LDA algorithms [36, 45, 55, 62, 69, 87] distribute massive data sets across multiple cores/processors to accelerate topic modeling. For example, parallel GS (PGS) [45] approximates the asynchronous GS sampling process by the synchronous updating of the global topic distributions. Similarly, FGS and SGS have been also implemented in the PGS parallel architecture. To reduce communication and synchronization costs in PGS [45, 70], a blackboard architecture has been proposed to deal with state synchronization for large clusters of workstations [55]. Parallel VB (PVB) [23, 87] uses the Map-Reduce paradigm to speed up LDA. Although these parallel algorithms can have almost the same topic modeling performance as batch counterparts, they require expensive parallel hardware; as a result the performance-to-price ratio remains unchanged. Moreover, parallel algorithms require additional resources to communicate and synchronize the topic distributions from distributed cores/processors, e.g., 1024 processors can achieve only around 700 times speedup with a parallel efficiency of approximately 0.7 [45]. Finally, it is difficult to write and debug parallel algorithms on distributed systems.

## 7.2.2  Residual Belief Propagation

Generally, there are two message-passing schedules in BP [57]. The first is the synchronous schedule $f^s$, which updates all messages (7.5) in parallel simultaneously at iteration $t$ based on the messages at previous iteration $t-1$:

$$f^s(\mu^{t-1}{}_{1,1}, \ldots, \mu^{t-1}_{W,D})$$
$$= \{f(\mu^{t-1}_{-1,-1}), \ldots, f(\mu^{t-1}_{-w,-d}) \ldots, f(\mu^{t-1}_{-W,-D})\}, \qquad (7.29)$$

where $f$ is the message update function (7.5) and $\mu_{-w,-d}$ is all set of messages excluding $\mu_{w,d}$. The second is the asynchronous schedule $f^a$, which updates the message of each topic label $z^k_{w,d}$ in a certain order, immediately influencing other neighboring message updates within the same iteration $t$:

$$f^a(\mu^{t-1}_{1,1}, \ldots, \mu^{t-1}_{W,D}) = \{\mu^t_{1,1}, \ldots, f(\mu^{t,t-1}_{-w,-d}), \ldots, \mu^{t-1}_{W,D}\}, \qquad (7.30)$$

where the message update equation $f$ is applied to each message one at a time in some order.

RBP [73] is an asynchronous BP that converges faster than synchronous BP. Suppose that the messages, $\mu^t = \{\mu^t_{1,1}, \ldots, \mu^t_{W,D}\}$, will converge to a set of fixed-points, $\mu^* = \{\mu^*_{1,1}, \ldots, \mu^*_{W,D}\}$, in the synchronous schedule (7.29). To speedup convergence in the asynchronous schedule (7.30), we choose to first update the message $\mu_{w,d}$ with the largest distance $\|\mu^t_{w,d} - \mu^*_{w,d}\|$ or $\|\mu^{t-1}_{w,d} - \mu^*_{w,d}\|$, which will efficiently influence its neighboring messages. However, we cannot directly measure the distance between a current message and its unknown fixed-point value.

**Fig. 7.9** RBP minimizes the
largest lower bound first

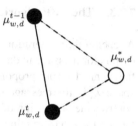

Alternatively, we can derive a lower bound on this distance that can be calculated easily. Using the triangle inequality, we get

$$\|\mu_{w,d}^{t} - \mu_{w,d}^{t-1}\| \leq \|\mu_{w,d}^{t} - \mu_{w,d}^{*}\| + \|\mu_{w,d}^{t-1} - \mu_{w,d}^{*}\|. \tag{7.31}$$

In this way, we first minimize the largest lower bound $\|\mu_{w,d}^{t} - \mu_{w,d}^{t-1}\|$ in Fig. 7.9, which defines the message residual [21],

$$r_{w,d}(k) = x_{w,d}\|\mu_{w,d}^{t}(k) - \mu_{w,d}^{t-1}(k)\|, \tag{7.32}$$

where $x_{w,d}$ is the number of word counts. The message residual $r_{w,d}(k) \to 0$ implies the convergence of RBP. More theoretical analysis on convergence property of RBP can be found in [21].

The computational cost of sorting (7.32) is expensive because the number of nonzero (NNZ) residuals $r_{w,d}$ is very large in the document-word matrix. In practice, we turn to sorting the accumulated residuals at vocabulary,

$$r_w = \sum_{k} r_w(k), \tag{7.33}$$

$$r_w(k) = \sum_{d} r_{w,d}(k), \tag{7.34}$$

which can be updated during message passing at a negligible computational cost. The time complexity of sorting (7.33) is at most $\mathcal{O}(W \log W)$. In many big corpora, the vocabulary size $W$ is a constant independent of the number of documents $D$.

### 7.2.3 Active Belief Propagation

In this section, we introduce ABP within the RBP framework. First, we describe in detail how to actively select the subset of documents and topics at each iteration for fast topic modeling. Second, we explain the effectiveness of ABP by Zipf's law. Finally, we discuss ABP's relationship to previous algorithms.

### 7.2.3.1  The ABP Algorithm

At each training iteration, ABP actively selects the subset of documents $\lambda_d D$ from corpus for message updating and passing, where the parameter $\lambda_d \in (0, 1]$ controls the scanned corpus proportion. For each document, ABP searches the subset of topic space $\lambda_k K$ for message updating and passing, where the parameter $\lambda_k \in (0, 1]$ controls the proportion of the searched topic space; that is, ideally ABP consumes a faction $(\lambda_d \lambda_k)$ of training time required by BP at each iteration.

Based on (7.32), we define the residuals at topics for specific documents as

$$r_d(k) = \sum_w r_{w,d}(k), \tag{7.35}$$

and accumulate all topic residuals at documents as

$$r_d = \sum_k r_d(k). \tag{7.36}$$

After each iteration, we sort $r_d(k)$ in descending order for all topics, and select the subset topics $\lambda_k K$ with the largest residuals $r_d(k)$ for each document. We also sort $r_d$ in descending order for all documents, and select the subset documents $\lambda_d D$ with the largest residuals $r_d$. In the following iteration, we update and pass only messages for the subset documents $\lambda_d D$ and the subset topics $\lambda_k K$, and thus save enormous training time in each loop. At a negligible computational cost, the sorted residuals can be updated during message-passing process. The computational cost of initial sorting (7.35) and (7.36) using the standard "quick sort" algorithm are $\mathcal{O}(K \log K)$ and $\mathcal{O}(D \log D)$. If the successive residuals are in almost sorted order, only a few swaps will restore the sorted order by the standard "insertion sort" algorithm, thereby saving lots of sorting time.[2]

For the selected $\lambda_k K$ topics, we need to normalize the local messages. At the first iteration $t = 1$, ABP runs the same as BP that updates and normalizes all messages for all topics. In the successive iterations $t \geq 2$, ABP actively selects the subset $\lambda_k K$ topics based on residuals for message updating and passing. We normalize the local messages for the selected topics $k \in \lambda_k K$ by

$$\hat{\mu}^t_{w,d}(k) = \frac{\mu^t_{w,d}(k)}{\sum_k \mu^t_{w,d}(k)} \times \sum_k \hat{\mu}^{t-1}_{w,d}(k), \, k \in \lambda_k K, \tag{7.37}$$

where $\hat{\mu}^{t-1}_{w,d}$ is the normalized message in the previous iteration, $\hat{\mu}^t_{w,d}(k)$ is the normalized message in the current iteration, and $\mu^t_{w,d}(k)$ is the unnormalized message updated according to (7.5). In this way, we need only $\lambda_k K$ iterations to avoid calculating the normalization factor $Z$ with $K$ iterations.

---

[2] In practice, partial sorting for $k$ largest elements is more efficient and retains almost the same performance.

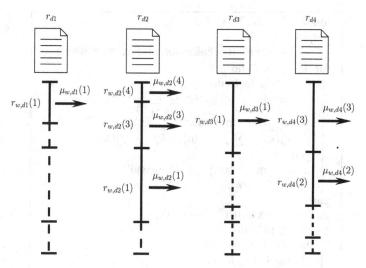

**Fig. 7.10** An example of active message passing

Figure 7.10 shows an example of the active message-passing process. The solid and dashed segments denote normalized messages in proportion. The solid segments and arrows denote those selected messages for updating and passing, while the dashed segments denote those unchanged messages. Suppose that documents $d1, d2, d3$, and $d4$ have largest residuals $r_{d1}, r_{d2}, r_{d3}$, and $r_{d4}$. We select these four documents in the subset $\lambda_d D$ for active message passing. For the document $d1$, only $r_{w,d1}(1)$ is in the subset $\lambda_k K$ so that the message $\mu_{w,d1}(1)$ is updated and passed. For the document $d2$, three residuals $r_{w,d2}(4), r_{w,d2}(3)$, and $r_{w,d2}(1)$ are ranked in the subset $\lambda_k K$, and thus the messages $\mu_{w,d2}(4), \mu_{w,d2}(3)$, and $\mu_{w,d2}(1)$ are updated and passed. Similar rules are applied to documents $d3$ and $d4$. Because only the subset of messages are updated and passed, ABP consumes significantly less computation than BP in each loop.

We dynamically refine and sort residuals at each iteration. Due to the contraction effect [21], the largest residuals in the previous iterations may be ranked lower in the subsequent iterations. As a result, those unchanged messages will be updated and passed as indicated by the dashed segments in Fig. 7.10. In this sense, ABP retains almost the same message information as BP.

Within the RBP framework [21], ABP will converge to a unique fixed-point by satisfying the following two assumptions:

1. The message update equation (7.5) is a contraction mapping $f$.
2. For every message $\mu_{w,d}(k)$, there is a finite time interval in which the message update (7.5) is executed at least once.

The first assumption is often satisfied in most existing message-passing algorithms based on factor graphs or clustered graphs [21]. With Fig. 7.10, we explain that the second assumption is often satisfied in practice due to the contraction effects.

**input**    : $\mathbf{x}$, $K$, $T$, $\alpha$, $\beta$, $\lambda_d$, $\lambda_k$.
**output**   : $\theta_d$, $\phi_w$.
$\mu_{w,d}^0(k) \leftarrow$ random initialization and normalization;
**for** $d \leftarrow 1$ **to** $D$ **do**
    **for** $k \leftarrow 1$ **to** $K$ **do**
        $\mu_{w,d}^1(k) \propto \dfrac{[\mu_{-w,d}^0(k)+\alpha] \times [\mu_{w,-d}^0(k)+\beta]}{\sum_w \mu_{w,-d}^0(k)+W\beta}$;
        $\mu_{w,d}^1(k) \leftarrow$ normalize$(\mu_{w,d}^1(k))$;
        $r_d^1(k) \leftarrow \sum_w x_{w,d} |\mu_{w,d}^1(k) - \mu_{w,d}^0(k)|$;
    **end**
    $\lambda_k K \leftarrow$ quick sort$(r_d^1(k),$ 'descend'$)$;
    $r_d^1 \leftarrow \sum_k r_d^1(k)$;
**end**
$\lambda_d D \leftarrow$ quick sort$(r_d^1,$ 'descend'$)$;
**for** $t \leftarrow 2$ **to** $T$ **do**
    **for** $d \in \lambda_d D$ **do**
        **for** $k \in \lambda_k K$ **do**
            $\mu_{w,d}^t(k) \propto \dfrac{[\mu_{-w,d}^{t-1}(k)+\alpha] \times [\mu_{w,-d}^{t-1}(k)+\beta]}{\sum_w \mu_{w,-d}^{t-1}(k)+W\beta}$;
            $\mu_{w,d}^t(k) \leftarrow \dfrac{\mu_{w,d}^t(k)}{\sum_k \mu_{w,d}^t(k)} \times \sum_k \mu_{w,d}^{t-1}(k)$;
            $r_d^t(k) \leftarrow \sum_w x_{w,d} |\mu_{w,d}^t(k) - \mu_{w,d}^{t-1}(k)|$;
        **end**
        $\lambda_k K \leftarrow$ insertionsort $(r_d^t(k),$ 'descend'$)$;
        $r_d^t \leftarrow \sum_k r_d^t(k)$;
    **end**
    $\lambda_d D \leftarrow$ insertionsort $(r_d^t,$ 'descend'$)$;
**end**
$\theta_d(k) \leftarrow [\boldsymbol{\mu}_{\cdot,d}^T(k)+\alpha]/\sum_k[\boldsymbol{\mu}_{\cdot,d}^T(k)+\alpha]$;
$\phi_w(k) \leftarrow [\boldsymbol{\mu}_{w,\cdot}^T(k)+\beta]/\sum_w[\boldsymbol{\mu}_{w,\cdot}^T(k)+\beta]$;

**Fig. 7.11**  The ABP algorithm

Figure 7.11 summarizes the ABP algorithm. When $t = 1$, ABP scans the entire corpus and searches the complete topic space, and in the meanwhile computes and sorts residuals by the "quick sort" algorithm. The purpose of this initial sweep is to calculate and store all residuals (7.35) and (7.36). For $2 \leq t \leq T$, ABP actively selects the subset documents $\lambda_d D$ and the subset topics $\lambda_k K$ for message updating and passing. At the end of each iteration, ABP dynamically refines and sorts residuals by the "insertion sort" algorithm. It terminates when the maximum number of iterations $T$ is reached or the convergence condition is satisfied.

The time complexity of ABP is $\mathcal{O}(\lambda_d \lambda_k K D T)$, where $K$ is the number of topics, $D$ the number of documents, and $T$ the number of iterations. The space complexity of ABP is $\mathcal{O}(KD)$, which is the same as the conventional BP or other batch LDA algorithms. Due to memory limitation for all batch LDA algorithms, it is hard to fit massive corpora with a large number of topics into the memory of a common desktop computer (for example, 3G RAM). We have two straightforward strategies to address the memory problem. First, we may integrate ABP with online stochastic optimization [10] or extend ABP on the blackboard parallel architecture [55]. Similar to other online and parallel LDA algorithms, online ABP requires a fixed memory requirement for each mini-batch data, while parallel ABP can use the large memory space from clusters of workstations. Second, we may combine ABP with the block optimization framework [71, 72] that sequentially reads blocks of data from the hard disk into memory for optimization. Nevertheless, these extensions of ABP are nontrivial at present, and to be studied in our future work such as [68, 77].

Here we show an alternative implementation of ABP. For massive corpora, initially sorting $r_d$ in (7.36) requires at most a computational complexity of $\mathcal{O}(D \log D)$. This cost is very high when $D$ is very large, for example, $D = 10^8$. Alternatively, we may define residuals at the fixed vocabulary like (7.33) and (7.34). We may sort $r_w$ with a significantly less computational cost of $\mathcal{O}(W \log W)$ because $W$ is often a fixed value whereas $D$ is a huge number for a massive corpora. In this case, ABP actively selects the subset of vocabulary words, $\lambda_w W$, $\lambda_w \in (0, 1]$, for topic modeling at each iteration. Due to page limitation of this chapter, we do not show the experimental results of this alternative implementation. Interested readers can find related source codes of this implementation in [86].

### 7.2.3.2 Document Residuals Follow Zipf's Law

The residual (7.32) reflects the magnitude of the gradient descent to maximize (7.4). The larger the residual the larger the gradient descent. Since ABP updates the selected subset of messages with the largest residuals, it in effect optimizes (7.4) using the largest gradient descents. According to the stochastic optimization theory [11], the subset of the gradient descents can perform well as the entire batch gradient descents when the data set is very large. Using the ENRON data set in Table 7.3 with $K = 10$, we show the log-log plot of the document residuals (7.36) relative to the rank of documents based on (7.36) in Fig. 7.12. Overall, the residuals are reduced with the increase of the training iterations $t \in \{100, 200, 300, 400, 500\}$, which indicates that the batch gradient descents are becoming smaller and smaller owing to convergence. Approximately, the residuals follow Zipf's law [90] during training,

$$\log(r_d) \approx a \log(rank) + b, \tag{7.38}$$

where $a$ and $b$ are constants. Zipf's law implies that small residuals are extremely common, whereas large residuals are extremely rare. More specifically, we find that top 20 % documents are responsible for almost 80 % residuals during training, which seems to follow the Pareto principle or the 80-20 rule,[3] stating that roughly 80 % of the effects come from 20 % of the causes. Therefore, ABP can efficiently use only 20 % documents to obtain 80 % gradient descents to optimize (7.4) at each training iteration, leading to a fast speed while achieving a comparable topic modeling

---

[3] http://en.wikipedia.org/wiki/Pareto_principle.

**Fig. 7.12** The residual
follows Zipf's law at
different iterations
$t \in \{100, 200, 300,$
$400, 500\}$ on the ENRON
data set

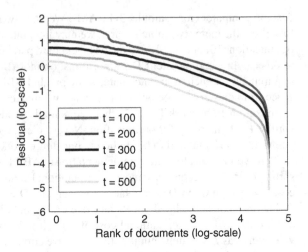

accuracy. Since many other natural data sets approximately follow Zipf's law, the
basic idea of ABP can be extended to speeding up other learning algorithms for these
data sets.

### 7.2.3.3 Relationship to Previous Algorithms

Unlike the parallel LDA algorithms [42, 45], ABP is a batch LDA algorithm that uses
only the single processor/core. It selects only a mini-batch of documents from the
entire corpus based on residuals for message updating and passing at each iteration.
From this perspective, ABP is similar to OVB [28] but with three main distinctions.
First, OVB sequentially read each mini-batch from the data stream without selection.
Second, OVB discards each mini-batch after one look. Finally, ABP can converge
to a local optimum of the BP's objective function, while OVB requires specific
parameter settings to achieve this goal. SOI [42], an extension of OVB, randomly
reuse previous seen mini-batches, so it does not schedule those mini-batches for a
better performance.

ABP differs from RVB [61] though they both use the residual-based dynamical
scheduling techniques. First, ABP is a batch algorithm based on the BP inference
while RVB is derived from the OVB [28] inference. Second, ABP uses a more efficient sorting method while RVB uses a relatively complicated sampling technique
for dynamical scheduling. Third, ABP can dynamically schedule documents, vocabulary words, and topics for the maximum speedup effects, while RVB can schedule
only mini-batches of documents. Finally, RVB uses only the lower bound of residuals (7.36), which may lower the scheduling efficiency.

The FGS and SGS algorithms [48, 70] are two important improvements over the
batch GS algorithm when $K$ is very large. The basic idea of FGS is to combine
both message update and sampling process together by introducing an upper bound

$\hat{Z}$ for the normalization factor $Z$. Similar to FGS, we may define an upper bound and pass only the largest segments $\mu_{w,d}(k)/\hat{Z}$ in BP to reduce the computation cost. However, this strategy does not work because we will not be able to update and pass those short segments (small messages) illustrated in Fig. 7.10, leading to serious loss of information. Although such a strategy is faster, its accuracy is much lower than BP. So, we formulate ABP within the RBP framework [21], which dynamically refines and sorts residuals to determine the best subset $\lambda_k K$ topic space at each iteration.

## 7.3 Type-2 Fuzzy Latent Dirichlet Allocation

Human action recognition is one of the challenging problems in computer vision. Potential applications for action recognition are many and include motion capture, human–computer interaction, biomedical analysis, surveillance and security, and sport and entertainment analysis, etc. There are several visual cues for recognizing human actions, e.g., motion [17] and shape [56]. Recently, one of the important motion cues for human action recognition is called "bag-of-words"(BOW) paradigm. The success of this paradigm has been largely attributed to its ability to encode statistics of local spatial–temporal interest point features. The typical BOW processing pipeline for object recognition problems in computer vision consists of three major steps. First, the local features of a collection of images are extracted by detectors [19]. Second, a codebook is constructed from high-dimensional local features by dimensionality reduction and clustering methods such as $k$-means [22] or k-medoids [62]. Finally, a probabilistic topic model (TM) [9] is built to describe the generative process of visual words. For example, Fig. 7.13a shows six frames corresponding to six actions: "boxing", "hand clapping", "hand waving", "jogging", "running", and "walking". The white box is the spatial–temporal cube containing interest point features extracted from each frame. In Fig. 7.13b, each cube is clustered into a "motion

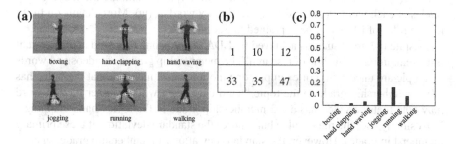

**Fig. 7.13** The "bag-of-words" (BOW) representation for human action recognition: **a** local feature extraction (each *white* box corresponds to a spatial–temporal interest point feature) from frames of six actions: boxing, hand clapping, hand waving, jogging, running, and walking. **b** Each spatial–temporal interest point is mapped to a "motion word" by dimensionality reduction and clustering methods (the number is the word index in the vocabulary). **c** Topic models assign an action topic label to each motion word, and thus a video sequence can be represented as the proportions of action topics using a histogram

word" index in the vocabulary. As a result, a video sequence can be viewed as a "document"containing a bag of motion words. In the rest of the chapter, we use "video" and "document" interchangeably. Note that "vocabulary word" means the unique "motion word" in the vocabulary. After TMs assign an action topic to each word, the video sequence can be represented as a mixture of action topics by a histogram in Fig. 7.13c. Although BOW loses some structural dependencies between motion words, it has been confirmed to be very effective in many object recognition tasks, such as natural scene categorization [22] and human action recognition [46, 62]. In this chapter, we propose type-2 fuzzy topic models (T2 FTM) to encode word semantic uncertainties for human action recognition based on the BOW paradigm.

Our models are motivated by the recent development of TMs such as latent Dirichlet allocation (LDA) [8], which has many important applications in computer vision such as scene recognition [22] and human action recognition [46, 62]. LDA is a hierarchical Bayesian model assuming that a video is a mixture of topics and a topic is a mixture of vocabulary words. It encourages different motion words within the same video to have the same topic assignment, and also encourages the same vocabulary word in different videos to have the same topic assignment. To avoid clustering all motion words into one topic, LDA penalizes the topics having so many motion words by a normalization factor to ensure that all topics contain almost the equal number of motion words. The best topic assignment is achieved by maximizing the joint probability of LDA, which results in three estimated topic-specific parameters: $\theta_k(d)$, $\phi_k(w)$ and $\mu_k(\{w, d\})$, where $k$ is the topic index, $d$ is the video index, and $w$ is the word index in the vocabulary. We may interpret these three parameters as type-1 fuzzy membership functions (MF). For example, $\theta_k(d)$ is the membership grade that the video $d$ belongs to the topic $k$, $\phi_k(w)$ is the membership grade that the vocabulary word $w$ belongs to the topic $k$, and $\mu_k(\{w, d\})$ is the membership grade that the motion word $\{w, d\}$ to the topic $k$. In this sense, LDA can be interpreted as a set of topic-specific IF-THEN rules, i.e., if the document $d$ and the vocabulary word $w$ are in topic $k$, then the motion word $\{w, d\}$ is also in topic $k$. Based on its membership grade $\mu_k(\{w, d\})$, each motion word will in turn vote to estimate the membership grades of each video $\theta_k(d)$ and each vocabulary word $\phi_k(w)$. More details on fuzzy if-then rules of LDA will be explained in Sect. 7.3.1.

Despite the great success achieved by LDA, there is one unsolved and important issue remaining in this line of research. The membership grades of videos and words to a topic are uncertain. For example, one motion word may vote that the video has a 0.1 membership grade to the topic "hand clapping", while another motion word may vote that the video has a 0.2 membership grade to the topic "hand clapping". LDA simply uses the mean value but ignores the standard deviation of these primary membership grades. However, the standard deviation of membership grades encode the higher-order uncertainty that has been confirmed useful in many real-world applications [41].

In this section, we attempt to address the above mentioned issue by combining type-2 fuzzy sets (T2 FS) [39] with topic models, referred to as type-2 fuzzy topic models (T2 FTMs). The MF of T2 FS is three-dimensional, where the primary membership grade describes uncertainty of objects to the FS, and the secondary membership grade evaluates the uncertainty of the primary membership grade itself. As far as the above example is concerned, we can use the secondary membership grades 0.8 and 0.4 to encode uncertainty of the video's primary grades 0.1 and 0.2 voted by two motion words to the topic "hand clapping". In T2 MF notations, we can use the T2 MF {0.8/0.1, 0.4/0.2} to encode the uncertainty of two motion words in a video. More clearly, we explicitly replace T1 MFs $\theta_k(d)$, $\phi_k(w)$ and $\mu_k(\{w, d\})$ by corresponding T2 MFs $\tilde{\theta}_k(d)$, $\tilde{\phi}_k(w)$ and $\tilde{\mu}_k(\{w, d\})$, which use the secondary MF to describe the uncertainty of primary MF. We believe that the additional secondary grades can improve the performance of human action categorization.

Following recent developments of T2 FSs [41], we implement two T2 FTMs. The first is the interval T2 FTM (IT2 FTM), which uses interval secondary MF assuming equal secondary grades [38]. We use the standard deviation of primary grades to quantify the length of interval. Therefore, the crisp value calculations in LDA will become the interval operations that can retain all uncertainties until topic assignment. The second is the vertical-slice T2 FTM (VT2 FTM), which uses the discrete unequal secondary grades to describe the uncertainty of each vertical slice of primary grades [41]. We calculate the centroid [31] of each vertical slice to infer the primary grade, where the uncertainty encoded by the vertical slice can be propagated by "meet" and "union" T2 FS operations [32]. For both T2 FTMs, we develop efficient message-passing algorithms to estimate parameters for action recognition. Experimental results on KTH [52] and Weizmann [7] human action data sets demonstrate that both implementations can yield state-of-the-art performance.

Integrating T2 FS [41] with probabilistic framework [6] has found many real-world applications such as classification of MPEG VBR video traffic [35], evaluation of welded structures [44], speech recognition [78, 79], handwritten Chinese character recognition [80–82], and classification of battlefield ground vehicles [67]. In the traditional Bayesian view, we can use the probabilistic distribution to quantify "randomness" uncertainties of signal features extracted from objects. The uncertainty of the parameters of probabilistic distributions can be further described by their conjugate prior distributions in order for tractable approximate inference. For example, LDA itself uses the multinomial distribution with parameter $\phi_k(w)$ to encode the likelihood of a vocabulary word given a topic $k$, and then uses the Dirichlet distribution $\text{Dir}(\phi_k|\beta)$ with hyperparameter $\beta$ to describe the uncertainty of the multinomial parameter $\phi_k(w)$. Generally speaking, Bayesian methods use a similar way as T2 FS to describe higher-order uncertainties but with one main distinction. The former requires approximate inference techniques such as Gibbs sampling, variational Bayes and belief propagation to estimate parameters of probabilistic distributions, while the latter uses T2 FS operations and type-reduction techniques [66] to make final decisions. In our view, combining Bayesian and T2 FS methods will shed more light on building powerful classification systems.

### 7.3.1 Topic Models

Figure 7.1a shows the original three-layer graphical representation of LDA [8]. It uses the document-specific topic proportion $\theta_d(k)$ to generate a topic label $z^k_{w,d,i} \in \{0, 1\}$, $\sum_{k=1}^{K} z^k_{w,d,i} = 1$, $1 \leq k \leq K$, which in turn generates each observed word token $i$ at index $w$ in the document $d$ based on the topic-specific multinomial distribution $\phi_k(w)$ over the vocabulary words. We use the shaded circles to represent the observed variables. Both multinomial parameters $\theta_d(k)$ and $\phi_k(w)$ are generated by two Dirichlet distributions with hyperparameters $\alpha$ and $\beta$, respectively. For simplicity, we consider only the smoothed LDA with fixed symmetric Dirichlet hyperparameters $\alpha$ and $\beta$ [26]. The plates indicate replications. For example, the document $d$ repeats $D$ times in the corpus, the word tokens $w_n$ repeats $N_d$ times in the document $d$, the vocabulary size is $W$, and there are $K$ topics. The joint probability of LDA [8] in Fig. 7.1 is

$$P(\mathbf{x}, \mathbf{z}, \boldsymbol{\theta}, \boldsymbol{\phi} | \alpha, \beta) \propto \prod_{k=1}^{K} P(\phi_k | \beta) \prod_{d=1}^{D} P(\theta_d | \alpha) \prod_{i=1}^{N_d} P(w_i | z^k_i, \phi_k) P(z^k_i | \theta_d). \quad (7.39)$$

Based on the Drichlet-Multinomial conjugacy, integrating out the multinomial parameters $\{\theta_d, \phi_k\}$ in (7.39) yields the joint probability of the collapsed LDA [26, 27],

$$p(\mathbf{x}, \mathbf{z}; \alpha, \beta) \propto \prod_d \prod_k \Gamma\left(\sum_w x_{w,d} z^k_{w,d} + \alpha\right) \prod_w \prod_k \Gamma\left(\sum_d x_{w,d} z^k_{w,d} + \beta\right)$$

$$\times \prod_k \Gamma\left(\sum_{w,d} x_{w,d} z^k_{w,d} + W\beta\right)^{-1}, \quad (7.40)$$

where $\Gamma(\cdot)$ is the gamma function and $x_{w,d} z^k_{w,d} = \sum_{i=1}^{x_{w,d}} z^k_{w,d,i}$ can recover the topic configuration over the word tokens in Fig. 7.1. The best topic labeling configuration $\mathbf{z}^*$ is obtained by maximizing (7.40) in terms of $\mathbf{z}$. The joint probability (7.40) can be represented by the factor graph [33], which facilitates the loopy belief propagation (BP) for approximate inference [76].

The BP algorithm is a message-passing algorithm that infers the conditional posterior probability called message $\mu_k(\{w, d\}) = p(z^k_{w,d}, x_{w,d} | \mathbf{z}_{-w,-d}, \mathbf{x}_{-w,-d})$. The message can be normalized efficiently using a local computation, i.e., $\sum_{k=1}^{K} \mu_k(\{w, d\}) = 1, 0 \leq \mu_k(\{w, d\}) \leq 1$. According to [76], the message can be updated by an iterative process. Given the inferred messages, the multinomial parameter matrices can be estimated by

$$\theta_k(d) = \sum_w x_{w,d} \mu_k(\{w, d\}) + \alpha, \quad (7.41)$$

$$\phi_k(w) = \frac{\sum_d x_{w,d} \mu_k(\{w, d\}) + \beta}{\sum_{w,d} x_{w,d} \mu_k(\{w, d\}) + W\beta}, \quad (7.42)$$

where $\phi_k(w)$ is normalized to satisfy $\sum_w \phi_k(w) = 1$. Based on the estimated parameter matrices, the message is updated as follows,

$$\mu_k(\{w, d\}) = \frac{\theta_k^{-w}(d)\phi_k^{-d}(w)}{\sum_k \theta_k^{-w}(d)\phi_k^{-d}(w)}, \quad (7.43)$$

where $-w$ and $-d$ denote excluding the current $x_{w,d}\mu_{w,d}(k)$ from the matrices (7.41) and (7.42), and $\mu_k(\{w, d\})$ is normalized locally to satisfy $\sum_k \mu_k(\{w, d\}) = 1$. After random initializing the messages, we repeat Eqs. (7.41)–(7.43) for $T$ iterations until convergence of the messages. In practice, $T \leq 300$ is enough to produce accurate estimation of multinomial parameters $\theta_k(d)$ and $\phi_k(w)$. Note that we need only pass messages for $x_{w,d} \neq 0$. Because $\mathbf{x}$ is a very sparse matrix, the message update equation (7.43) is computationally fast by sweeping all nonzero elements in the sparse matrix $\mathbf{x}$.

### 7.3.1.1 Fuzzy IF-THEN Rules for LDA

Although LDA is a probabilistic model, it can be interpreted as a set of fuzzy IF-THEN rules—a machinery which is employed in almost all applications of fuzzy logic. We assume that a topic is a fuzzy set, where the membership functions $\theta_k(d)$ and $\phi_k(w)$ describe the membership grades of the document $d$ and word $w$ belonging to the topic $k$, respectively. For a rule base of $1 \leq k \leq K$ rules, each having two antecedents, the $k$th rule is

$$\text{Topic } k : \text{ IF } d \in \theta_k(d) \text{ and } w \in \phi_k(w), \text{ THEN } \{w, d\} \in \mu_k(\{w, d\}). \quad (7.44)$$

Using the product $t$-norm operation in fuzzy logic, we obtain the same message update equation (7.43). Note that we need to normalize the membership grade $\mu_k(\{w, d\})$ in terms of $k$ to satisfy $\sum_k \mu_k(\{w, d\}) = 1$.

Figure 7.14 shows one of the $K$ IF-THEN rules of LDA. Figure 7.14a is the membership grade of a document belonging to the $k$th topic, i.e., $\theta_k(d)$. Figure 7.14b

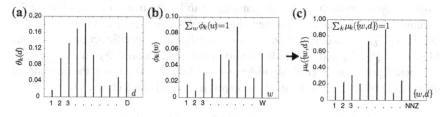

**Fig. 7.14** The fuzzy IF-THEN rules of LDA: **a** the membership function $\theta_k(d)$ over $1 \leq d \leq D$ documents, **b** the membership function $\phi_k(w)$ over $1 \leq w \leq W$ vocabulary words, and **c** the inferred membership function $\mu_k(\{w, d\})$ over the number of nonzero ($NNZ$) elements in the document-word matrix

is the membership grade of a vocabulary word belonging to the $k$th topic, i.e., $\phi_k(w)$. According to the rule (7.44), we can infer the membership grade of a motion word belonging to the $k$th topic, i.e., $\mu_k(\{w, d\})$. The inferred membership grade $\mu_k(\{w, d\})$ is in turn used to update membership functions $\theta_k(d)$ and $\phi_k(w)$ according to (7.41) and (7.42), respectively. After several iterations, all membership grades will reach the fixed stationary points [76]. The convergence is ensured by the expectation–maximization algorithm of LDA [76]. Intuitively, the rule is correct because if a video belongs to the topic "hand clapping", and a vocabulary word also belongs to the topic "hand clapping", it is likely that this motion word in this video belongs to the topic "hand clapping".

The transformation of a probabilistic model to a set of fuzzy IF-THEN rules shows that both methods can account for uncertainties in observed data but from different perspectives. The probabilistic method tries to use the inferred model distribution to approximate the true distribution of the observed data by minimizing the Kullback–Leibler (KL) divergence between these two distributions according to the maximum-likelihood estimation. In contrast, the fuzzy IF-THEN rules clearly show the intrinsic relations among different FSs represented by MFs, which are readily understood by users. Both methods have their own advantages to represent the same real-world problem.

### 7.3.1.2 Labeled LDA (L-LDA)

LDA is an unsupervised learning method because the $K$ topics do not correspond to object classes. Although the unsupervised LDA can also produce good results in human action recognition [46], the labeled LDA (L-LDA) is more suitable for pattern classification tasks [50, 62]. Figure. 7.15 shows the graphical representation of L-LDA. Compared with LDA in Fig. 7.1, the membership grade $\theta_k$ is constrained to be the class label $\Lambda_k$ of the video. L-LDA constrains that each video is only associated with its action label set $\Lambda_k(d)$. For example, if the video has three types of motions "boxing", "hand clapping", and "hand waving", there are only three types of topic labels assigned to this video corresponding to three types of motions. More specifically, the constraint is $\Lambda_k(d) = \{1, 1, 1, 0, 0, 0\}, 1 \leq k \leq 6$. The parameter estimation of labeled LDA is almost the same with Eqs. (7.41)–(7.43), but the topic labels assigned to each video are constrained to be the motion classes during the training process. Given the observed video-word matrix $\mathbf{x}_{W \times D}$ and the action label matrix $\Lambda_{K \times D} = \{\Lambda_k(d)\}$, L-LDA aims to infer the video-specific action proportions $\Theta_{K \times D} = \{\theta_k(d)\}$ and action distribution over visual words $\Phi_{K \times W} = \{\phi_k(w)\}$. The belief propagation (BP) algorithm for L-LDA [76, 86] infers the conditional distribution of action labels called *messages*, $\mu_k(\{w, d\})$, as follows,

$$\mu_k(\{w, d\}) = \frac{\theta_k^{-w}(d)\phi_k^{-d}(w)\Lambda_k(d)}{\sum_k \theta_k^{-w}(d)\phi_k^{-d}(w)\Lambda_k(d)}, \tag{7.45}$$

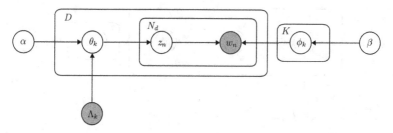

**Fig. 7.15** The graphical representation of labeled LDA (L-LDA) [50, 62]. L-LDA is more suitable for categorization tasks because it constrains the membership grade $\theta_k$ by $\Lambda_k$, which corresponds to the observed class labels in the training video sequences

which is normalized locally to satisfy $\sum_{k=1}^{K} \mu_k(\{w, d\}) = 1$. With the observed action labels $\Lambda_k$ for each video, the parameter estimation of $\theta_k$ and $\phi_k$ in Eqs. (7.41) and (7.42) will be more reasonable, accurate, and faster. In the recognition process, the unlabled video sequences are assumed to have all topic labels without constraint. We assign the class label to the video sequence with the highest topic proportion. More specifically, for the test video $d$, we fix $\phi_k$ and iterates Eqs. (7.43) and (7.41) without the $\Lambda_k$ constraint. Then, we classify this video into the action category by $k = \arg \max_k \theta_k(d)$.

### 7.3.2 Type-2 Fuzzy Topic Models (T2 FTMs)

Although topic models such as LDA and L-LDA gain great successes in text mining and computer vision, they use only mean value of messages of the motion words $\{w, d\}$ to estimate membership functions $\theta_k$ and $\phi_k$. As different words may play different roles in different documents, the variation of these messages should be considered to account for the complexity of human actions. Such parameter uncertainties of L-LDA can be described by T2 FS [40]. Previous studies also combined T2 FS with other graphical models for parameter uncertainties such as Gaussian mixture models [84], hidden Markov models [83], and Markov random fields [85].

Figure 7.16 shows the graphical representation for type-2 fuzzy topic models (T2 FTMs). The major difference from Fig. 7.26 is that the membership functions $\theta_k(d)$ and $\phi_k(w)$ become type-2 membership functions (T2 MFs) $\tilde{\theta}_k(d)$ and $\tilde{\phi}_k(d)$. Note that T2 MFs can use the secondary membership functions to describe the uncertainties of the primary memberships. Similar to the fuzzy IF-THEN rule (7.44), we interpret Fig. 7.16 by the T2 fuzzy IF-THEN rules. For a rule base of $1 \leq k \leq K$ rules, each having two antecedents, the $k$th rule is

$$\text{Topic } k : \text{ IF } d \in \tilde{\theta}_k(d) \text{ and } w \in \tilde{\phi}_k(w), \text{ THEN } \{w, d\} \in \tilde{\mu}_k(\{w, d\}). \quad (7.46)$$

In this chapter, we propose two implementations of T2 FTMs. The first is the interval T2 FTM, where the secondary MF is an interval. The second is the vertical-slice

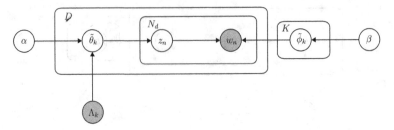

**Fig. 7.16** The graphical representation of type-2 fuzzy topic models, where $\tilde{\theta}_k$ and $\tilde{\phi}_k$ are type-2 fuzzy membership functions

T2 FTM, where the secondary MF is a histogram of weights. We shall develop efficient message-passing algorithms to estimate parameters in T2 FTMs.

### 7.3.2.1 Interval Type-2 Fuzzy Topic Model (IT2 FTM)

An example of the T2 MF $\tilde{\theta}_k(d)$ is illustrated in Fig. 7.17. We see that the primary memberships are uncertain in Fig. 7.17a, which can be described by either Gaussian shape or interval shape secondary memberships in Fig. 7.17b, c, respectively. The basic idea of IT2 FTM is to describe the T1 MF $\theta_k$ and $\phi_k$ by the corresponding interval

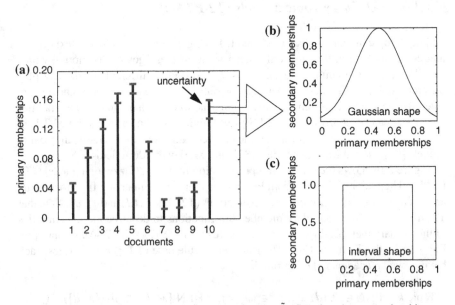

**Fig. 7.17** An example of type-2 fuzzy membership function $\tilde{\theta}_k(d)$: **a** the primary memberships are uncertain denoted by *red error bars*, **b** the uncertainty of primary memberships can be described by the Gaussian shape secondary membership grades, and **c** the uncertainty of primary memberships can be also described by the interval shape secondary membership grades

type-2 fuzzy MF: $\tilde{\theta}_k = [\underline{\theta}_k, \overline{\theta}_k]$ and $\tilde{\phi}_k = [\underline{\phi}_k, \overline{\phi}_k]$ as shown in Fig. 7.17c. The length of the interval is derived from the Gaussian shape MF in Fig. 7.17b. The reason why we do not directly use Gaussian shape secondary MF is that the T2 FS operation is intractable. So, we adopt the tractable interval calculations. In this way, we use the interval to represent the parameter uncertainty. Such "fuzziness" uncertainty will be passed by interval calculations until decision-making. In the following, we will derive the interval of parameters from message uncertainty in (7.41) and (7.42).

The video $d$ contains multiple motion word messages in Eq. (7.41). Unlike directly using the weighted sum in Eq. (7.41), we calculate the mean and standard deviation of these messages,

$$m_k(d) = \frac{\sum_w x_{w,d} \mu_k(\{w, d\})}{\sum_w x_{w,d}}, \tag{7.47}$$

$$\sigma_k(d) = \sqrt{\frac{\sum_w x_{w,d} [\mu_k(\{w, d\}) - m_k(d)]^2}{\sum_w x_{w,d}}}. \tag{7.48}$$

Equation (7.48) is the standard deviation of messages that reflects message uncertainty in the video $d$, which has been largely ignored in both LDA and L-LDA. Similarly, the messages of the motion word $w$ in the vocabulary occurs in many documents in Eq. (7.42). We also calculate the mean and standard deviation of these messages as follows,

$$m_k(w) = \frac{\sum_d x_{w,d} \mu_{w,d}(k)}{\sum_d x_{w,d}}, \tag{7.49}$$

$$\sigma_k(w) = \sqrt{\frac{\sum_d x_{w,d} [\mu_k(\{w, d\}) - m_k(w)]^2}{\sum_d x_{w,d}}}. \tag{7.50}$$

Equation (7.50) reflects message uncertainty of the vocabulary word $w$ across different videos, which has been also ignored in both LDA and L-LDA.

Due to the message uncertainties, we replace the messages in (7.41) and (7.42) by intervals, $\tilde{\mu}_k(d) = [\underline{\mu}_k(d), \overline{\mu}_k(d)]$ and $\tilde{\mu}_k(w) = [\underline{\mu}_k(w), \overline{\mu}_k(w)]$, where

$$\underline{\mu}_k(d) = \sum_w x_{w,d} [m_k(d) - u\sigma_k(d)], \tag{7.51}$$

$$\overline{\mu}_k(d) = \sum_w x_{w,d} [m_k(d) + u\sigma_k(d)], \tag{7.52}$$

$$\underline{\mu}_k(w) = \sum_d x_{w,d} [m_k(w) - u\sigma_k(w)], \tag{7.53}$$

$$\overline{\mu}_k(w) = \sum_d x_{w,d} [m_k(w) + u\sigma_k(w)], \tag{7.54}$$

where the parameter $u \in [0, +\infty)$ controls the length of intervals. The larger $u$ corresponds to more uncertainty of messages. When $u = 0$, the interval messages $\tilde{\mu}_k(d)$ and $\tilde{\mu}_k(w)$ reduce to crisp messages in Eqs. (7.41) and (7.42), so that IT2 FTM in Fig. 7.16 becomes standard L-LDA in Fig. 7.15. By the interval representation, we assume that all messages variate around the mean value within a certain standard deviation. To avoid the negative values in messages, we constrain all intervals in the range $[0, +\infty)$. If the left point of an interval is less than zero, we set the left point as zero. Although this strategy will lose some information, it works well in the later experiments. In practice, we use $0 \leq u \leq 3$ because it is unlikely that the message varies beyond the interval $[m - 3\sigma, m + 3\sigma]$.

Based on the message intervals, we may rewrite Eqs. (7.41) and (7.42) as

$$\tilde{\theta}_k(d) = \tilde{\mu}_k(d) + \alpha, \qquad (7.55)$$

$$\tilde{\phi}_k(w) = \frac{\tilde{\mu}_k(w) + \beta}{\sum_w \tilde{\mu}_k(w) + W\beta}, \qquad (7.56)$$

where the sum operation on intervals is the sum of intervals' left-end and right-end points. Notice that (7.56) involves interval normalization or division, having intervals in both numerator and denominator. We adopt the normalization method for interval weights [53], which assumes that the sum of normalized interval weights should be an interval centered around 1 with a minimal width, i.e., the midpoint is $\phi_k(w) = \frac{1}{2}[\underline{\phi}_k(w) + \overline{\phi}_k(w)]$ and it satisfies $\sum_w \phi_k(w) \approx 1$. Based on (7.55) and (7.56), we rewrite the message update equation (7.45) as

$$\tilde{\mu}_k(\{w, d\}) = \frac{\tilde{\theta}_k(d)\tilde{\phi}_k(w)\Lambda_k(d)}{\sum_k \tilde{\theta}_k(d)\tilde{\phi}_k(w)\Lambda_k(d)}, \qquad (7.57)$$

where the sum and product operations on interval sets are the sum and product calculation on the intervals' left-end and right-end points [40]. Again, Eq. (7.57) involves interval normalization similar to (7.56). To repeat Eqs. (7.51)–(7.54), we defuzzify the interval $\tilde{\mu}_{w,d}(k) = [\underline{\mu}_k(\{w, d\}), \overline{\mu}_k(\{w, d\})]$ by its midpoint,

$$\mu_k(\{w, d\}) = \frac{1}{2} \times [\underline{\mu}_k(\{w, d\}) + \overline{\mu}_k(\{w, d\})]. \qquad (7.58)$$

Combining Eqs. (7.55)–(7.57) yields the message-passing algorithm for IT2 FTM in Fig. 7.18, where $T$ is the number of iterations. First, we randomly initialize and normalize the membership grades $\mu_k(\{w, d\})$ for all messages (line 1). Second, we calculate the mean and standard deviation from messages and use intervals to describe uncertainty of membership functions $\tilde{\theta}_k(d)$ and $\tilde{\phi}_k(w)$ (lines 3–12), respectively. Finally, we reestimate and defuzzify the interval message $\tilde{\mu}_k(\{w, d\})$ using its mid-point. For the test video $d$, we fix $\tilde{\phi}_k(w)$ and run Eqs. (7.51)–(7.54) in Fig. 7.18 without the $\Lambda_k(d)$ constraint. Then, we classify this video into the action category

---

**input** : $\mathbf{x}$, $\Lambda$, $T$, $\alpha$, $\beta$, $u$.
**output** : $\tilde{\theta}_k(d)$, $\tilde{\phi}_k(w)$.
1 $\mu_k(\{w, d\}) \leftarrow$ random initialization and normalization;
2 **for** $t \leftarrow 1$ **to** $T$ **do**
3 $\quad m_k(d) \leftarrow \frac{\sum_w x_{w,d} \mu_k(\{w,d\})}{\sum_w x_{w,d}}$;
4 $\quad \sigma_k(d) \leftarrow \sqrt{\frac{\sum_w x_{w,d}[\mu_k(\{w,d\}) - m_k(d)]^2}{\sum_w x_{w,d}}}$;
5 $\quad m_k(w) \leftarrow \frac{\sum_d x_{w,d} \mu_{w,d}(k)}{\sum_d x_{w,d}}$;
6 $\quad \sigma_k(w) \leftarrow \sqrt{\frac{\sum_d x_{w,d}[\mu_k(\{w,d\}) - m_k(w)]^2}{\sum_d x_{w,d}}}$;
7 $\quad \underline{\mu}_k(d) \leftarrow \sum_w x_{w,d}[m_k(d) - u\sigma_k(d)]$;
8 $\quad \overline{\mu}_k(d) \leftarrow \sum_w x_{w,d}[m_k(d) + u\sigma_k(d)]$;
9 $\quad \underline{\mu}_k(w) \leftarrow \sum_d x_{w,d}[m_k(w) - u\sigma_k(w)]$;
10 $\quad \overline{\mu}_k(w) \leftarrow \sum_d x_{w,d}[m_k(w) + u\sigma_k(w)]$;
11 $\quad \tilde{\theta}_k(d) \leftarrow \tilde{\mu}_k(d) + \alpha$;
12 $\quad \tilde{\phi}_k(w) \leftarrow \frac{\tilde{\mu}_k(w) + \beta}{\sum_w \tilde{\mu}_k(w) + W\beta}$;
13 $\quad \tilde{\mu}_k(\{w, d\}) \leftarrow \frac{\tilde{\theta}_k(d)\tilde{\phi}_k(w)\Lambda_k(d)}{\sum_k \tilde{\theta}_k(d)\tilde{\phi}_k(w)\Lambda_k(d)}$;
14 $\quad \mu_k(\{w, d\}) \leftarrow \frac{1}{2} \times [\underline{\mu}_k(\{w,d\}) + \overline{\mu}_k(\{w,d\})]$;
15 **end**

---

**Fig. 7.18** The message-passing algorithm for IT2 FTM

by the maximum midpoint of the interval $\tilde{\theta}_k(d)$, i.e., $k = \arg \max_k \frac{1}{2}[\underline{\theta}_k(d) + \overline{\theta}_k(d)]$. The time complexity of Fig. 7.18 is slightly higher than that of L-LDA because at each iteration we need to update the standard deviation of messages (lines 4 and 6).

### 7.3.2.2 Vertical-Slice Type-2 Fuzzy Topic Model (VT2 FTM)

IT2 FTM describes uncertainty of primary memberships using continuous intervals, while VT2 FTM describes uncertainty of primary memberships using discrete histogram of secondary membership grades. The basic idea is inspired by (7.41) and (7.42), where the document membership or vocabulary word membership is an aggregation of motion word messages. We can imagine that these messages compose the discrete primary membership grades in each vertical slice of the T2 FSs $\tilde{\theta}_k(d)$ and $\tilde{\phi}_k(w)$. Figure 7.19a shows an example of VT2 MF $\tilde{\theta}_k(d)$, where the discrete primary membership grades are messages of motion words in (7.41). The uncertainty of the discrete primary membership grades are encoded by the histogram of secondary membership grades in Fig. 7.19b.

Suppose that we have two vertical slices (each containing two primary grades $\{a, b\}$ or $\{c, d\}$) from two T2 FSs, $\tilde{\theta}_k(d) = \{\frac{f(a)}{a}, \frac{f(b)}{b}\}$ and $\tilde{\phi}_k(w) = \{\frac{f(c)}{c}, \frac{f(d)}{d}\}$, where $f(\cdot)$ is the corresponding secondary membership grades. The *meet* operation

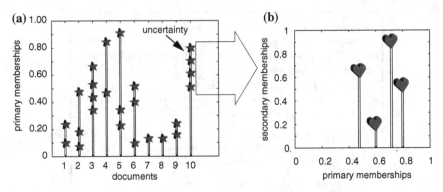

**Fig. 7.19** An example of vertical-slice T2 MF $\tilde{\theta}_k(d)$. **a** The primary memberships $\tilde{\theta}_k(d)$ are composed of discrete memberships of motion words $\mu_k(\{w, d\})$ within a video $d$. Each *star* represents the membership grade voted by a motion word. **b** A histogram of secondary membership grades describe the uncertainty of four primary memberships denoted by *heart*. We can type-reduce each vertical slice of multiple primary memberships by the weighted average operation to centroid of primary membership

[41] (or intersection) on these two vertical slices in rule (7.46) is

$$\tilde{\mu}_k(\{w, d\}) = \tilde{\theta}_k(d) \sqcap \tilde{\phi}_k(w) = \left\{ \frac{f(a)f(c)}{ac}, \frac{f(a)f(d)}{ad}, \frac{f(b)f(c)}{bc}, \frac{f(b)f(d)}{bd} \right\},$$
(7.59)

where the centroid of $\tilde{\mu}_k(\{w, d\})$ is

$$\text{Centroid}_{\tilde{\mu}_k(\{w,d\})} = \frac{f(a)f(c)ac + f(a)f(d)ad + f(b)f(c)bc + f(b)f(d)bd}{f(a)f(c) + f(a)f(d) + f(b)f(c) + f(b)f(d)}.$$
(7.60)

However, the *meet* operation (7.59) is a computationally expensive combinatorial problem. In practice, the number of primary grades on a vertical slice of $\tilde{\theta}_k(d)$ is equal to the number words within a document, which is often a large number for the combinatorial problem (7.59). To save computation, we first type-reduce $\tilde{\theta}_k(d)$ and $\tilde{\phi}_k(w)$ for each vertical slice in order to obtain the centroids $\theta_k(d)$ and $\phi_k(w)$,

$$\theta_k(d) = \frac{f(a)a + f(b)b}{f(a) + f(b)},$$
(7.61)

$$\phi_k(w) = \frac{f(c)c + f(d)d}{f(c) + f(d)}.$$
(7.62)

Then we update the message based on the centroids of $\tilde{\theta}_k(d)$ and $\tilde{\phi}_k(w)$ according to (7.43),

$$\mu_k(\{w, d\}) = \frac{f(a)a + f(b)b}{f(a) + f(b)} \times \frac{f(c)c + f(d)d}{f(c) + f(d)}, \qquad (7.63)$$

which can produce the same message centroid as (7.60), but can avoid the combinatorial problem in the *meet* operation (7.59).

The following question is how to design proper secondary grades to describe the uncertainty of primary memberships or motion messages in Fig. 7.19b. As far as $\tilde{\theta}_k(d)$ is concerned, different motion words within a video may play different roles quantified by $f_w(k)$. For $\tilde{\phi}_k(w)$, the same vocabulary word in different videos may have different importance weights $f_{w,d}$. According to (7.61) and (7.62), we rewrite (7.41) and (7.42) as

$$\theta_k(d) = \sum_w x_{w,d}\mu_k(\{w, d\})f_w(k) + \alpha, \qquad (7.64)$$

$$\phi_k(w) = \frac{\sum_d x_{w,d}\mu_k(\{w, d\})f_{w,d} + \beta}{\sum_{w,d} x_{w,d}\mu_k(\{w, d\})f_{w,d} + W\beta}, \qquad (7.65)$$

where $f_w(k)$ and $f_{w,d}$ are secondary grades as shown in Fig. 7.19b. We design $f_w(k)$ as

$$f_w(k) = \frac{N_{w,k}}{N_{w,-k}}, \qquad (7.66)$$

where $N_{w,k}$ is the number of vocabulary word $w$ in the topic $k$, and $N_{w,-k}$ is the number of vocabulary word $w$ out of the topic $k$. The idea of using the inverse word frequency to reduce the weighting of commonly occurring words is known to be effective, which captures the intuition that commonly occurring words in all topics are less useful in identifying the topic of a frame than the rarely occurring ones. Note that (7.66) is a prefixed value from the training set without tuning during the learning process. We design $f_{w,d}$ as the sparseness of $\mu_k(\{w, d\})$ in (7.65) with respect to $k$ as follows [30],

$$f_{w,d} = \frac{\sqrt{K} - (\sum_k \mu_k(\{w, d\}))/\sqrt{\sum_k \mu_k^2(\{w, d\})}}{\sqrt{K} - 1}. \qquad (7.67)$$

If the vector $\mu_k(\{w, d\})$ in terms of $k$ is sparse like $[0, 0, 0, 0, 0, 1]$, then $f_{w,d} = 1$. If the vector $\mu_k(\{w, d\})$ is smooth like $[1/6, 1/6, 1/6, 1/6, 1/6, 1/6]$, then $f_{w,d} = 0$. So, the weight (7.67) reflects the intuition that if a message is more focused on one topic, it will have a higher impact to the topic distribution over words (7.65). Note that (7.67) needs to be updated at each learning iteration.

**input** : $\mathbf{x}, \Lambda, T, \alpha, \beta, u.$
**output** : $\tilde{\theta}_k(d), \tilde{\phi}_k(w).$
1 $\mu_k(\{w, d\}) \leftarrow$ random initialization and normalization;
2 $f_w(k) = \frac{N_{w,k}}{N_{w,-k}};$
3 **for** $t \leftarrow 1$ **to** $T$ **do**
4 $\quad f_{w,d} = \frac{\sqrt{K} - (\sum_k \mu_k(\{w,d\}))/\sqrt{\sum_k \mu_k^2(\{w,d\})}}{\sqrt{K}-1};$
5 $\quad \theta_k(d) = \sum_w x_{w,d}\mu_k(\{w,d\})f_w(k) + \alpha;$
6 $\quad \phi_k(w) = \frac{\sum_d x_{w,d}\mu_k(\{w,d\})f_{w,d}+\beta}{\sum_{w,d} x_{w,d}\mu_k(\{w,d\})f_{w,d}+W\beta};$
7 $\quad \mu_k(\{w,d\}) = \frac{\theta_k^{-w}(d)\phi_k^{-d}(w)\Lambda_k(d)}{\sum_k \theta_k^{-w}(d)\phi_k^{-d}(w)\Lambda_k(d)};$
8 **end**

**Fig. 7.20** The message-passing algorithm for VT2 FTM

Figure 7.20 shows the message-passing algorithm for VT2 FTM. The basic difference lies in that we use secondary grades $f_w(k)$ and $f_{w,d}$ to update two primary memberships $\theta_k(d)$ and $\phi_k(w)$ in (7.64) and (7.65) (lines 5–6). The computational complexity is almost the same with that of L-LDA except that we need to additionally update the secondary grades (7.67) at each iteration (line 4). Indeed, the design of secondary functions (7.66) and (7.67) is arbitrary, if these designed functions can reflect our prior knowledge on specific tasks. For a test video $d$, we fix $\phi_k(w)$ and use Fig. 7.20 to infer the video-specific topic proportion $\theta_k(d)$. Then, we assign the class label to the video with the maximum topic proportion, i.e., $k = \arg\max_k \theta_k(d)$.

## 7.4 Topic Modeling Performance

### 7.4.1 Belief Propagation

We use four large-scale document data sets:

1. CORA [37] contains abstracts from the CORA research chapter search engine in machine learning area, where the documents can be classified into 7 major categories.
2. MEDL [88] contains abstracts from the MEDLINE biomedical chapter search engine, where the documents fall broadly into 4 categories.
3. NIPS [48] includes chapters from the conference "Neural Information Processing Systems", where all chapters are grouped into 13 categories. NIPS has no citation link information.
4. BLOG [20] contains a collection of political blogs on the subject of American politics in the year 2008. where all blogs can be broadly classified into 6 categories. BLOG has no author information.

**Table 7.2** Summarization of four document data sets

| Data sets | $D$ | $A$ | $W$ | $C$ | $N_d$ | $W_d$ |
|---|---|---|---|---|---|---|
| CORA | 2410 | 2480 | 2961 | 8651 | 57 | 43 |
| MEDL | 2317 | 8906 | 8918 | 1168 | 104 | 66 |
| NIPS | 1740 | 2037 | 13649 | – | 1323 | 536 |
| BLOG | 5177 | – | 33574 | 1549 | 217 | 149 |

Table 7.2 summarizes the statistics of four data sets, where $D$ is the total number of documents, $A$ is the total number of authors, $W$ is the vocabulary size, $C$ is the total number of links between documents, $N_d$ is the average number of words per document, and $W_d$ is the average vocabulary size per document.

### 7.4.1.1  BP for LDA

We compare BP with two commonly used LDA learning algorithms such as VB [8] (Here, we use Blei's implementation of digamma functions)[4] and GS [26].[5] under the same fixed hyperparameters $\alpha = \beta = 0.01$. We use MATLAB C/C++ MEX-implementations for all these algorithms, and carry out experiments on a common PC with CPU 2.4 GHz and RAM 4G. With the goal of repeatability, we have made our source codes and data sets publicly available [86][6]

To examine the convergence property of BP, we use the entire data set as the training set, and calculate the training perplexity [8] at every 10 iterations in the total of 1,000 training iterations from the same initialization. Figure 7.21 shows that the training perplexity of BP generally decreases rapidly as the number of training iterations increases. In our experiments, BP on average converges with the number of training iterations $T \approx 170$ when the difference of training perplexity between

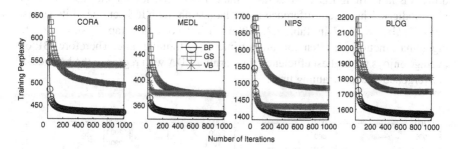

**Fig. 7.21** Training perplexity as a function of number of iterations when $K = 50$ on CORA, MEDL, NIPS, and BLOG

---

[4] http://www.cs.princeton.edu/~blei/lda-c/index.html.

[5] http://psiexp.ss.uci.edu/research/programs_data/toolbox.htm.

[6] https://mloss.org/software/view/399/.

two successive iterations is less than one. Further analysis reveals that BP on average uses more number of training iterations until convergence than VB ($T \approx 100$) but much less number of training iterations than GS ($T \approx 300$) on the four data sets. The fast convergence rate is a desirable property as far as the online [28] and distributed topic modeling for large-scale corpus are concerned.

The predictive perplexity for the unseen test set is computed as follows [2, 8]. To ensure all algorithms to achieve the local optimum, we use the 1,000 training iterations to estimate $\phi$ on the training set from the same initialization. In practice, this number of training iterations is large enough for convergence of all algorithms in Fig. 7.21. We randomly partition each document in the test set into 90 % and 10 % subsets. We use 1,000 iterations of learning algorithms to estimate $\theta$ from the same initialization while holding $\phi$ fixed on the 90 % subset, and then calculate the predictive perplexity on the left 10 % subset,

$$\mathscr{P} = \exp\left\{ -\frac{\sum_{w,d} x_{w,d}^{10\%} \log\left[\sum_k \theta_d(k)\phi_w(k)\right]}{\sum_{w,d} x_{w,d}^{10\%}} \right\}, \qquad (7.68)$$

where $x_{w,d}^{10\%}$ denotes word counts in the 10 % subset. Notice that the perplexity (7.68) is based on the marginal probability of the word topic label $\mu_{w,d}(k)$ in (7.19).

Figure 7.22 shows the predictive perplexity (average $\pm$ standard deviation) from fivefold cross-validation for different topics, where the lower perplexity indicates the better generalization ability for the unseen test set. Consistently, BP has the lowest perplexity for different topics on four data sets, which confirms its effectiveness for learning LDA. On average, BP lowers around 11 % than VB and 6 % than GS in perplexity. Figure 7.23 shows that BP uses less training time than both VB and GS. For a better illustration, we show only 0.3 times of the real training time of VB because of time-consuming digamma functions. In fact, VB runs as fast as BP if we remove digamma functions. So, we believe that it is the digamma functions that slow down VB in learning LDA. BP is faster than GS because it computes messages for word indices. The speed difference is the largest on the NIPS set due to its largest ratio $N_d/W_d = 2.47$ in Table 7.2. Although VB converges rapidly attributed to digamma functions, it often consumes triple more training time. Therefore, BP on average enjoys the highest efficiency for learning LDA with regard to the balance of convergence rate and training time.

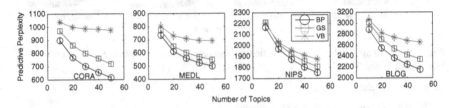

**Fig. 7.22** Predictive perplexity as a function of number of topics on CORA, MEDL, NIPS, and BLOG

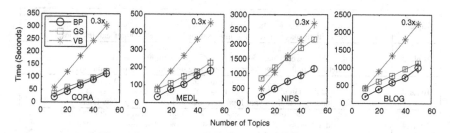

**Fig. 7.23** Training time as a function of number of topics on CORA, MEDL, NIPS, and BLOG. For VB, it shows 0.3 times of the real learning time denoted by 0.3x

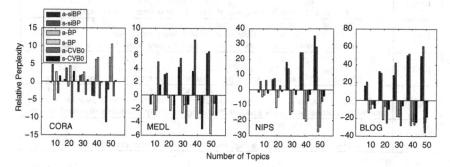

**Fig. 7.24** Relative predictive perplexity as a function of number of topics on CORA, MEDL, NIPS, and BLOG

We also compare six BP implementations such as siBP, BP, and CVB0 [2] using both synchronous and asynchronous update schedules. We name three synchronous implementations as s-BP, s-siBP, and s-CVB0, and three asynchronous implementations as a-BP, a-siBP, and a-CVB0. Because these six belief propagation implementations produce comparable perplexity, we show the relative perplexity that subtracts the mean value of six implementations in Fig. 7.24. Overall, the asynchronous schedule gives slightly lower perplexity than synchronous schedule because it passes messages faster and more efficiently. Except on CORA set, siBP generally provides the highest perplexity because it introduces subtle biases in computing messages at each iteration. The biased message will be propagated and accumulated leading to inaccurate parameter estimation. Although the proposed BP achieves lower perplexity than CVB0 on NIPS set, both of them work comparably well on other sets. But BP is much faster because it computes messages over word indices. The comparable results also confirm our assumption that topic modeling can be efficiently performed on word indices instead of tokens.

To measure the interpretability of a topic model, the *word intrusion* and *topic intrusion* are proposed to involve subjective judgements [14]. The basic idea is to ask volunteer subjects to identify the number of word intruders in the topic as well as the topic intruders in the document, where intruders are defined as inconsistent

| Topic 1 | recognition image system images network training speech figure model set image images figure object feature features recognition space visual distance recognitioni mage images training system speech setf eaturef eatures word |
|---------|---|
| Topic 2 | network networks units input neural learning hidden output training unit network networks neural input units output learning hidden layer weights network units networks input output neural hiddenl earning unitl ayer |
| Topic 3 | analog chip circuit figure neural input output time system network analog circuit chip output figure signal input neural time system analog neural circuitc hipfi gure output input times ystems ignal |
| Topic 4 | time model neurons neuron spike synaptic activity input firing information model neurons cells neuron cell visual activity response input stimulus neurons timem odeln eurons ynaptic spike cell activity input firing |
| Topic 5 | model visual figure cells motion direction input spatial field orientation time noise dynamics order results point model system values figure modelv isualfi gurem otionfi eldd irections patial cells imageo rientation |
| Topic 6 | function functions algorithm linear neural matrix learning space networks data function functions number set algorithm theorem tree bound learning class functionf unctions algorithms et theorem linear numberv ector cases pace |
| Topic 7 | learning network error neural training networks time function weight model function learning error algorithm training linear vector data set space learning errorn euraln etwork functionw eight training networks timeg radient |
| Topic 8 | data training set error algorithm learning function class examples classification training data set performance classification recognition test class error speech training data sete rrorl earning performance test neural numberc lassification |
| Topic 9 | model data models distribution probability parameters gaussian algorithm likelihood mixture model data models distribution gaussian probability parameters likelihood mixture algorithm modeld atad istributionm odels gaussian algorithmp robability parameters likelihood mixture |
| Topic 10 | learning state time control function policy reinforcement action algorithm optimal learning state control time model policy action reinforcement system states learning state timec ontrol policyf unction actionr einforcementa lgorithmm odel |

**Fig. 7.25** Top ten words of $K = 10$ topics of VB (*first line*), GS (*second line*), and BP (*third line*) on NIPS

words or topics based on prior knowledge of subjects. Figure 7.25 shows the top ten words of $K = 10$ topics inferred by VB, GS, and BP algorithms on NIPS set. We find no obvious difference with respect to word intrusions in each topic. Most topics share the similar top ten words but with different ranking orders. Despite significant perplexity difference, the topics extracted by three algorithms remains almost the same interpretability at least for the top ten words. This result coincides with [14] that the lower perplexity may not enhance interpretability of inferred topics. Similar phenomenon has also been observed in MRF-based image labeling problems [57]. Different MRF inference algorithms such as graph cuts and BP often yield comparable results. Although one inference method may find more optimal MRF solutions, it does not necessarily translate into better performance compared to the ground truth. The underlying hypothesis is that the ground truth labeling configuration is often less optimal than solutions produced by inference algorithms. For example, if we manually label the topics for a corpus, the final perplexity is often higher than that of solutions returned by VB, GS, and BP. For each document, LDA provides the equal number of topics $K$ but the ground truth often uses the unequal number of topics to explain the observed words, which may be another reason why the overall perplexity of learned LDA is often lower than that of the ground truth.

To test this hypothesis, we compare the perplexity of labeled LDA (L-LDA) [50] with LDA in Fig. 7.26. L-LDA is a supervised LDA that restricts the hidden topics

**Fig. 7.26**  Perplexity of
L-LDA and LDA on four
data sets

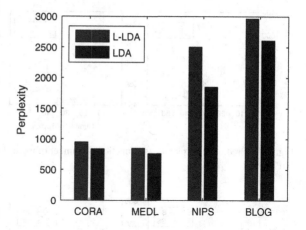

as the observed class labels of each document. When a document has multiple class labels, L-LDA automatically assigns one of the class labels to each word index. In this way, L-LDA resembles the process of manual topic labeling by human, and its solution can be viewed as close to the ground truth. For a fair comparison, we set the number of topics $K = 7, 4, 13, 6$ of LDA for CORA, MEDL, NIPS, and BLOG according to the number of document categories in each set. Both L-LDA and LDA are trained by BP using 500 iterations from the same initialization. Figure 7.26 confirms that L-LDA produces higher perplexity than LDA, which partly supports that the ground truth often yields the higher perplexity than the optimal solutions of LDA inferred by BP.

The underlying reason may be that the three topic modeling rules encoded by LDA are still too simple to capture human behaviors in finding topics. Under this situation, improving the formulation of topic models such as LDA is better than improving inference algorithms to enhance the topic modeling performance significantly. Although the proper settings of hyperparameters can make the predictive perplexity comparable for all state-of-the-art approximate inference algorithms [2], we still advocate BP because it is faster and more accurate than both VB and GS, even if they all can provide comparable perplexity and interpretability under the proper settings of hyperparameters.

### 7.4.1.2  BP for ATM and RTM

The GS algorithm for learning ATM is implemented in the MATLAB topic modeling toolbox.[7] We compare BP and GS for learning ATM based on 500 iterations on training data. Figure 7.27 shows the predictive perplexity (average ± standard deviation) from fivefold cross-validation. On average, BP lowers 12 % perplexity than GS, which is consistent with Fig. 7.22. Another possible reason for such significant

---

[7] http://psiexp.ss.uci.edu/research/programs_data/toolbox.htm.

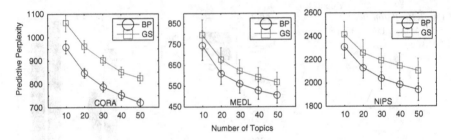

**Fig. 7.27** Predictive perplexity as a function of number of topics for ATM on CORA, MEDL, and NIPS

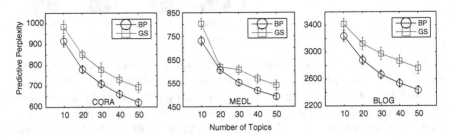

**Fig. 7.28** Predictive perplexity as a function of number of topics for RTM on CORA, MEDL, and BLOG

improvements may be attributed to our assumption that all coauthors of the document account for the word topic label using multinomial instead of uniform probabilities.

The GS algorithm for learning RTM is implemented in the R package.[8] We compare BP with GS for learning RTM using the same 500 iterations on training data set. Based on the training perplexity, we manually set the weight $\xi = 0.15$ in Fig. 7.8 to achieve the overall superior performance on four data sets.

Figure 7.28 shows predictive perplexity (average $\pm$ standard deviation) on fivefold cross-validation. On average, BP lowers 6 % perplexity than GS. Because the original RTM learned by GS is inflexible to balance information from different sources, it has slightly higher perplexity than LDA (Fig. 7.22). To circumvent this problem, we introduce the weight $\xi$ in (7.28) to balance two types of messages, so that the learned RTM gains lower perplexity than LDA.

We also examine the link prediction performance of RTM. We define the link prediction as a binary classification problem. As with [16], we use the Hadmard product of a pair of document topic proportions as the link feature, and train an SVM [15] to decide if there is a link between them. Notice that the original RTM [16] learned by the GS algorithm uses the GLM to predict links. Figure 7.29 compares the F-measure (average $\pm$ standard deviation) of link prediction on fivefold cross-validation. Encouragingly, BP provides significantly 15 % higher F-measure over

---

[8] http://cran.r-project.org/web/packages/lda/.

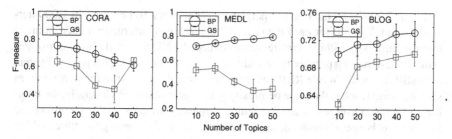

**Fig. 7.29** F-measure of link prediction as a function of number of topics on CORA, MEDL, and BLOG

GS on average. These results confirm the effectiveness of BP for capturing accurate topic structural dependencies in document networks.

## 7.4.2 Residual Belief Propagation

We carry out experiments on six publicly available data sets: (1) 20 newsgroups (NG20),[9] (2) BLOG [20], (3) CORA [37], (4) MEDLINE [89], (5) NIPS [25], and (6) WEBKB.[10] Table 7.3 summarizes the statistics of six data sets, where $D$ is the total number of documents in the corpus, $W$ is the number of words in the vocabulary, $N_d$ is the average number of word tokens per document, and $W_d$ is the average number of word indices per document. All subsequent figures show results on six data sets in the above order. We compare RBP with three state-of-the-art approximate inference methods for LDA including VB [8], GS [26], and sBP [76] under the same fixed hyperparameters $\alpha = \beta = 0.01$. We use MATLAB C/C++ MEX-implementations for all these algorithms [86], and carry out the experiments on a desktop computer with CPU 2.4 GHz and RAM 4G.

**Table 7.3** Statistics of six document data sets

| Data set | $D$ | $W$ | $N_d$ | $W_d$ |
|---|---|---|---|---|
| NG20 | 7505 | 61188 | 239 | 129 |
| BLOG | 5177 | 33574 | 217 | 149 |
| CORA | 2410 | 2961 | 57 | 43 |
| MEDLINE | 2317 | 8918 | 104 | 66 |
| NIPS | 1740 | 13649 | 1323 | 536 |
| WEBKB | 7061 | 2785 | 50 | 29 |

---

[9] http://people.csail.mit.edu/jrennie/20Newsgroups.

[10] http://csmining.org/index.php/webkb.html.

Figure 7.30 shows the training perplexity [2] at every 10 iterations in 1,000 iterations when $K = 10$ for each data set. All algorithms converge to a fixed-point of training perplexity within 1,000 iterations. Except the NIPS set, VB always converges at the highest training perplexity. In addition, GS converges at a higher perplexity than both sBP and RBP. While RBP converge at almost the same training perplexity as sBP, it always reaches the same perplexity value faster than sBP. Generally, the training algorithm converges when the training perplexity difference, at two consecutive iterations, is below a threshold. In this chapter, we set the convergence threshold to 1 because the training perplexity decreases very little after this threshold is satisfied in Fig. 7.30.

Figure 7.31 illustrates the number of training iterations until convergence on each data set for different topics $K \in \{10, 20, 30, 40, 50\}$. The number of iterations until convergence seems insensitive to the number of topics. On the BLOG, CORA, and WEBKB sets, VB uses the minimum number of iterations until convergence, consistent with the previous results in [76]. For all data sets, GS consumes the maximum number of iterations until convergence. Unlike the deterministic message updating in VB, sBP, and RBP, GS uses the stochastic message updating scheme accounting for the largest number of iterations until convergence. Although sBP costs the less number of iterations until convergence than GS, it still uses the much more number of iterations than VB. By contrast, through the informed dynamic scheduling for asynchronous message passing, RBP on average converges more rapidly than sBP for all data sets. In particular, on the NG20, MEDLINE, and NIPS sets, RBP on average uses a comparable or even less number of iterations than VB until convergence.

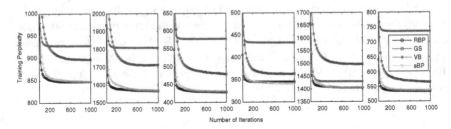

**Fig. 7.30**  Training perplexity as a function of the number of iterations when $K = 10$

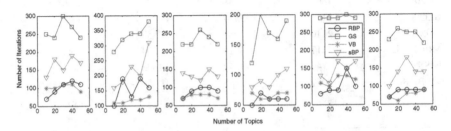

**Fig. 7.31**  The number of training iterations until convergence as a function of number of topics

Figure 7.32 shows the training time in seconds until convergence on each data set for different topics $K \in \{10, 20, 30, 40, 50\}$. Surprisingly, while VB usually uses the minimum number of iterations until convergence, it often consumes the longest training time for these iterations. The major reason may be attributed to the time-consuming digamma functions in VB, which takes at least triple more time for each iteration than GS and sBP. If VB removes the digamma functions, it runs as fast as sBP. Because RBP uses a significantly less number of iterations until convergence than GS and sBP, it consumes the least training time until convergence for all data sets in Fig. 7.32.

We also examine the predictive perplexity of all algorithms until convergence based on a tenfold cross-validation. The predictive perplexity for the unseen test set is computed as that in [2]. Figure 7.33 shows the box plot of predictive perplexity for tenfold cross-validation when $K = 50$. The plot produces a separate box for ten predictive perplexity values of each algorithm. On each box, the central mark is the median, the edges of the box are the 25th and 75th percentiles, the whiskers extend to the most extreme data points not considered outliers, and outliers are plotted individually by the red plus sign. Obviously, VB yields the highest predictive perplexity, corresponding to the worst generalization ability. GS has a much lower predictive perplexity than VB, but it has a much higher perplexity than both sBP and RBP. The underlying reason is that GS samples a topic label from the messages without retaining all possible uncertainties. The residual-based scheduling scheme of RBP not only speeds up the convergence rate of sBP, but also slightly lowers the predictive perplexity. The reason is that RBP updates fast-convergent messages to efficiently influence those slow-convergent messages, reaching fast to the local minimum of the predictive perplexity.

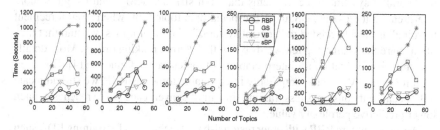

**Fig. 7.32** Training time until convergence as a function of number of topics

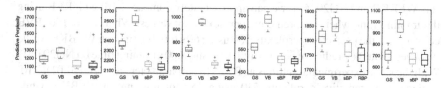

**Fig. 7.33** Predictive perplexity for tenfold cross-validation when $K = 50$

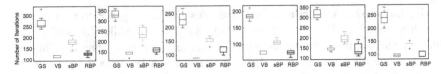

**Fig. 7.34** The number of training iterations until convergence for tenfold cross-validation when $K = 50$

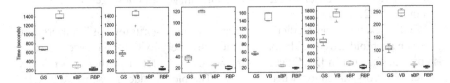

**Fig. 7.35** Training time until convergence for tenfold cross-validation when $K = 50$

Figures 7.34 and 7.35 illustrate the box plots for the number of iterations and the training time until convergence for tenfold cross-validation when $K = 50$. Consistent with Figs. 7.31 and 7.32, VB consumes the minimum number of iterations, but has the longest training time until convergence. GS has the maximum number of iterations, but has the second longest training time until convergence. Because RBP improves the convergence rate over sBP, it consumes the least training time until convergence.

To measure the interpretability of inferred topics, Fig. 7.36 shows the top ten words of each topic when $K = 10$ on CORA set using 500 training iterations. We observe that both sBP and RBP can infer almost the same topics as other algorithms except the topic one, where sBP identifies the "pattern recognition" topic but RBP infers the "parallel system" topic. It seems that both sBP and RBP obtain slightly more interpretable topics than GS and VB especially in topic four, where "reinforcement learning" is closely related to "control systems". For other topics, we find that they often share the similar top ten words but with different ranking orders. More details on subjective evaluation for interpretability of topics can be found in [14]. However, even if GS and VB yield comparably interpretable topics as RBP, we still advocate RBP because it consumes less training time until convergence while reaches a much lower predictive perplexity value.

We also compare RBP with other residual-based techniques for training LDA such as RVB [60, 61]. It is not easy to make a fair comparison because RBP is an offline learning but RVB is an online learning algorithm. However, using the same data sets WEBKB and NG20 [61], we can approximately compare RBP with RVB using the training time when the predictive perplexity converges. When $K = 100$, RVB converges at the predictive perplexity 600 using 60 s training time on WEBKB, while it converges at the predictive perplexity 1050 using 600 s training time on NG20 [60]. With the same experimental settings as RVB (hyperparameters $\alpha = \beta = 0.01$), RBP achieves the predictive perplexity 540 using 35 s for training on WEBKB, while it achieves the predictive perplexity 1004 using 420 s for training on NG20. The significant speedup is because RVB involves relatively slower digamma function

| | |
|---|---|
| Topic 1 | network recognition system training neural word set speech input classifier<br>network recognition word system speech training set neural classifier model<br>**classifier recognition word set training classification speech model system class**<br>recognition training classifier set network word data classification speech system |
| Topic 2 | circuit neuron chip signal input network output system analog neural<br>signal circuit system chip neural analog output network input current<br>**circuit neural chip analog system input network output current voltage**<br>signal circuit chip analog system output neural input current sound |
| Topic 3 | network function learning neural weight algorithm input error result linear<br>network function learning weight neural input output unit error equation<br>**network unit input weight output training neural learning layer hidden**<br>network unit input weight output learning neural training layer hidden |
| Topic 4 | neuron model input cell network spike synaptic firing system pattern<br>neuron model cell input spike synaptic activity firing response pattern<br>**neuron model input cell synaptic spike network activity firing system**<br>neuron model cell input spike synaptic activity network firing response |
| Topic 5 | learning function algorithm action policy problem control optimal step reinforcement<br>learning algorithm function action problem result policy step theorem states<br>**learning control action system model task policy reinforcement step dynamic**<br>model control system learning movement controller robot motor position task |
| Topic 6 | network unit learning weight input output training hidden set neural<br>network unit learning input training weight output hidden task set<br>**function algorithm learning point problem linear number order result case**<br>function algorithm learning problem result action number bound policy set |
| Topic 7 | model data parameter function algorithm distribution gaussian set network vector<br>model data distribution parameter gaussian algorithm probability component method density<br>**model data distribution probability parameter gaussian method mean noise density**<br>model data distribution parameter method gaussian function set algorithm probability |
| Topic 8 | image visual model field motion cell direction object map images<br>image visual field object images motion map direction feature cell<br>**visual model field motion cell image direction object map eye**<br>image visual field motion cell images object map model direction |
| Topic 9 | model control system network learning neural movement input motor position<br>model control system learning movement robot controller motor position eye<br>**signal frequency component filter sound system auditory information analysis data**<br>set model algorithm structure problem representation rules graph vector cluster |
| Topic 10 | data model set error training algorithm network function method learning<br>data set training error function algorithm vector method problem network<br>**vector data feature set image problem features point representation algorithm**<br>function learning point algorithm network vector linear matrix equation order |

**Fig. 7.36** Top ten words of $K = 10$ topics for GS (*blue*), VB (*red*), sBP (*green*), and RBP (*black*) on CORA set

computations, and adopts a more complicated sampling method based on residual distributions for dynamic scheduling.

## 7.4.3 Active Belief Propagation

Our experiments aim to verify the accelerating effects of ABP compared with that of other state-of-the-art batch LDA algorithms including VB [8], GS [26], FGS [48], SGS [69], CVB0 [2], and synchronous BP [76]. We use four publicly available document data sets [48]: NIPS, ENRON, NYTIMES, and PUBMED. Previous studies [48] have revealed that the speedup effect is relatively insensitive to the number of documents in the corpus. Because of the memory limitation for batch LDA algorithms, we randomly select 15,000 documents from the original NYTIMES data set, and 80,000

**Table 7.4** Statistics of four document data sets

| Data sets | $D$ | $W$ | $N_d$ | $W_d$ |
| --- | --- | --- | --- | --- |
| NIPS | 1500 | 12419 | 311.8 | 217.6 |
| ENRON | 39861 | 28102 | 160.9 | 93.1 |
| NYTIMES | 15000 | 84258 | 328.7 | 230.2 |
| PUBMED | 80000 | 76878 | 68.4 | 46.7 |

documents from the original PUBMED data set for experiments. Table 7.4 summarizes the statistics of four data sets, where $D$ is the total number of documents in the corpus, $W$ is the number of words in the vocabulary, $N_d$ is the average number of word tokens per document, and $W_d$ is the average number of word indices per document.

We randomly partition each data set into halves with one for training set and the other for test set. We calculate the training perplexity [2] on the training set after 500 iterations as follows,

$$\mathcal{P} = \exp\left\{ -\frac{\sum_{w,d} x_{w,d} \log\left[\sum_k \theta_d(k)\phi_w(k)\right]}{\sum_{w,d} x_{w,d}} \right\}. \tag{7.69}$$

Usually, the training perplexity will decrease with the increase of the number of training iterations. The algorithm often converges if the change of training perplexity at successive iterations is less than a predefined threshold. In our experiments, we set the threshold to one because the decrease of training perplexity will be very small after satisfying this threshold.

The predictive perplexity for the unseen test set is computed as follows [2]. On the training set, we estimate $\phi$ from the same random initialization after 500 iterations. For the test set, we randomly partition each document into 80 % and 20 % subsets. Fixing $\phi$, we estimate $\theta$ on the 80 % subset by the training algorithms from the same random initialization after 500 iterations, and then calculate the predictive perplexity on the rest 20 % subset,

$$\mathcal{P} = \exp\left\{ -\frac{\sum_{w,d} x_{w,d}^{20\%} \log\left[\sum_k \theta_d(k)\phi_w(k)\right]}{\sum_{w,d} x_{w,d}^{20\%}} \right\}, \tag{7.70}$$

where $x_{w,d}^{20\%}$ denotes word counts in the the 20 % subset. The lower predictive perplexity represents a better generalization ability.

For all data sets, we fix the same hyperparameters $\alpha = 2/K$ and $\beta = 0.01$ [48]. The CPU time per iteration is measured after sweeping the entire data set. We report the average CPU time per iteration after $T = 500$ iterations, which practically ensures that GS and FGS converge in terms of training perplexity. For a fair comparison, we use the same random initialization to examine all algorithms with 500 iterations. We implement all algorithms based on the MEX C++/MATLAB/Octave platform, where GS/FGS C++ source codes are the same as those in [48], and SGS source codes are the same as those in Mallet package.[11] To repeat our experiments, we have made all source codes and data sets publicly available [86]. Without the parallel implementations, all algorithms are run on the Sun Fire X4270 M2 server with two 6-core 3.46 GHz CPUs and 128 GB RAMs.

### 7.4.3.1 Parameters $\lambda_d$ and $\lambda_k$

First, we examine two parameters $\lambda_d$ and $\lambda_k$ in ABP on the relatively smaller NIPS data set, because we were wondering how these two parameters would influence the topic modeling accuracy of ABP. The parameter $\lambda_d \in (0, 1]$ controls the proportion of documents to be scanned, and the parameter $\lambda_k \in (0, 1]$ controls the proportion of topics to be searched at each iteration. The smaller values correspond to the faster speed of ABP. We choose the training perplexity $\mathscr{P}_{\lambda_d=1,\lambda_k=1}$ of ABP with 500 iterations when $\lambda_d = \lambda_k = 1$ as the benchmark. The relative training perplexity is the difference between this benchmark and the ABP's training perplexity (7.69) with 500 iterations at other parameter values,

$$\Delta\mathscr{P} = \mathscr{P}_{\lambda_d,\lambda_k} - \mathscr{P}_{\lambda_d=1,\lambda_k=1}. \tag{7.71}$$

**Fig. 7.37** The relative training perplexity as a function of $K$ when the parameter **a** $\lambda_d \in \{0.1, 0.2, 0.3, 0.4, 0.5\}$ and **b** $\lambda_k \in \{0.1, 0.2, 0.3, 0.4, 0.5\}$

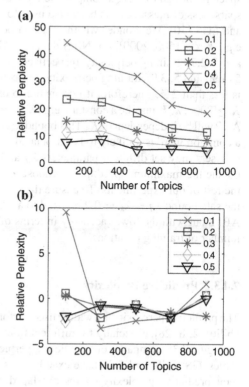

---

[11] http://mallet.cs.umass.edu/.

Fixing $\lambda_k = 1$, we change $\lambda_d$ from 0.1 to 0.5 with a step size 0.1. Figure 7.37a shows the relative training perplexity as a function of $K$ when $\lambda_d \in \{0.1, 0.2, 0.3, 0.4, 0.5\}$. We see that the relative perplexity decreases when $\lambda_d$ increases, which confirms the fact that scanning more documents at each iteration will yield the lower training perplexity under different topics $K$. Notice that the relative perplexity decreases as $K$ increases. This phenomenon shows that when $K$ is very large, even scanning a small portion of documents is enough to provide a comparable topic modeling accuracy. When $\lambda_d \geq 0.2$ we also see that the relative perplexity is less than 20, which in practice is a negligible difference on the common document data set. Therefore, $\lambda_d = 0.2$ is a safe bound to guarantee a good topic modeling accuracy with a relatively faster speed. Although a smaller $\lambda_d < 0.2$ will produce a even faster speed, it will cause an obvious degradation of the topic modeling accuracy in terms of perplexity with a fixed number of training iterations.

Fixing $\lambda_d = 1$, we also change $\lambda_k$ from 0.1 to 0.5. Figure 7.37b shows the relative training perplexity as a function of $K$ when $\lambda_k \in \{0.1, 0.2, 0.3, 0.4, 0.5\}$. Surprisingly, there is no big difference when $\lambda_k = 0.1$ and $\lambda_k = 0.5$ especially when $K \geq 300$. This phenomenon implies that only a small proportion of topics plays a major role when $K$ is very large. When $\lambda_k \leq 0.5$, ABP achieves even a lower perplexity value than ABP with $\lambda_k = 1$. The reason is that most documents have very sparse messages when $K$ is very large, and thus searching the subset of topic space is enough to yield a comparable topic modeling accuracy. Such a property as sparseness of messages has been also used to speedup topic modeling by FGS [48] and SGS [70]. We wonder whether $\lambda_k$ can be even smaller when $K$ is very large, e.g., $K \in \{1500, 2000\}$. On NIPS data set, ABP with $\lambda_k = 0.05$ achieves 555.89 and 542.70 training perplexity, respectively. In contrast, ABP with $\lambda_k = 1$ achieves 543.90 and 533.97 training perplexity, respectively. The relative training perplexity is less than 2 %. Therefore, it reasonable to expect that when $K$ is very large like $K \geq 2000$, $\lambda_k K$ may be a constant, e.g., $\lambda_k K = 100$. In this case, the training time of ABP will be independent of $K$. This bound $\lambda_k K = 100$ is reasonable because usually a common word is unlikely to be associated with more than 100 topics in practice.

Users may set different parameters $\lambda_d$ and $\lambda_k$ for different speedup effects. To pursue the maximum speedup, we choose $\lambda_d = \lambda_k = 0.1$ referred to as ABP1 in the rest of our experiments. To ensure the topic modeling accuracy, we also choose the safe bound $\lambda_d = \lambda_k = 0.2$ referred to as ABP2. Obviously, ABP1 is faster than ABP2 but with the lower accuracy in terms of the training perplexity using a fixed number of training iterations.

### 7.4.3.2 Predictive Perplexity

The predictive perplexity is a widely used performance measure for the generalization ability [2, 8, 28], especially for different training algorithms of LDA.

Figure 7.38 compares the predictive perplexity as a function of the number of topics. GS, FGS, and SGS have exactly the same topic modeling accuracy, so that their predictive perplexity curves overlap denoted by blue squares. On ENRON,

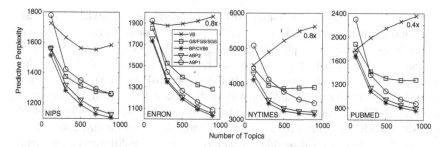

**Fig. 7.38** Predictive perplexity as a function of topics $K = \{100, 300, 500, 700, 900\}$ on NIPS, ENRON, NYTIMES, and PUBMED data sets. The notation $0.8x$ denotes the predictive perplexity is multiplied by $0.8$

NYTIMES, and PUBMED data sets, VB shows an obvious overfitting phenomenon, where the predictive perplexity increases with the number of topics $K$. Also, VB often yields the highest predictive perplexity. For a better visualization, we multiply VB's predictive perplexity value by $0.8$ and $0.4$ on ENRON, NYTIMES, and PUBMED data sets, respectively. Confronted with a large number of topics on large data sets, VB often generalizes badly to predict unseen test set. The major reason is that VB aims to optimize an approximate variational distribution with a gap from the joint distribution (7.4). Our experiments show that this gap or bias is obvious in case of the massive data set containing a large number of topics. In practice, VB may choose the proper hyperparameters to correct this bias [2]. On all data sets, BP consistently yields the lowest predictive perplexity. This result is consistent with [76]. Unlike VB, BP decreases the predictive perplexity with the increase of the number of topics $K$ without overfitting. Because BP directly infers the conditional posterior probability from (7.4), it often generalizes to unseen documents well in practice. Similarly, CVB0 infers the messages on the word tokens instead of the word indices, so it has almost the same predictive perplexity as BP denoted by the red star curve as shown in Fig. 7.38.

With $\lambda_d = \lambda_k = 0.2$, ABP2 performs almost the same as BP/CVB0, especially on ENRON and PUBMED data sets. This result shows that scanning 20 % documents and searching 20 % topics at each iteration are enough to provide a comparable topic modeling accuracy as BP/CVB0. With $\lambda_d = \lambda_k = 0.1$, ABP1 provides a relatively higher predictive perplexity than BP/CVB0. On the relatively smaller data set NIPS, ABP1 performs even worse than VB when $K = 100$. However, on the relatively larger data set PUBMED and when $K$ is large, e.g., $K = 900$, ABP1 shows a comparable predictive perplexity as BP/CVB0. When compared with GS/FGS/SGS, ABP1 is worse under different topics on the relatively smaller NIPS set, but is much better on the relatively larger data sets: ENRON, NYTIMES, and PUBMED, when $K$ is large. In conclusion, ABP2 has a comparable performance as BP/CVB0, and outperforms GS/FGS/SGS and VB on all data sets. While ABP1 has a higher predictive perplexity than ABP2 and BP/CVB0 on all data sets, it often outperforms GS/FGS/SGS and VB for larger data sets when the number of topics is very large ($K \geq 500$).

With the same training iterations, ABP1 has a relatively higher predictive perplexity than BP, which implies that ABP1 loses some topic modeling accuracy to achieve a significant speedup. This phenomenon resembles the lossy/lossless compression for images. To obtain a higher compression ratio, lossy techniques will generally discard some detailed image information. In this sense, ABP can be viewed as a lossy learning algorithm for topic modeling with the fast speed. By contrast, FGS and SGS are lossless GS algorithms for topic modeling. When $\lambda_d = \lambda_k = 1$, ABP becomes the lossless RBP algorithm for accurate topic modeling with the fast convergence speed [73]. In different real-world applications, we may have different requirements for speed and accuracy. Users may choose different parameters $\{\lambda_d, \lambda_k\}$ to trade off between the topic modeling speed and accuracy.

### 7.4.3.3 Speedup Effects

Figure 7.39 shows the CPU time (second) per iteration as a function of $K$ for VB, GS, FGS, SGS, BP, CVB0, ABP1, and ABP2. The training time of all algorithms increases linearly with $K$ except FGS on the PUBMED data set. When $K$ is very large, FGS [48] will locate the targeted topics more efficiently so that its training time does not increase linearly with $K$. Because FGS does not visit all possible topic space to sample the targeted topic, it is much faster than GS when $K$ is very large. On four data sets, SGS is much faster than FGS because it avoids almost all surplus computations during the sampling process. Similar to GS, both FGS and SGS still scan the entire corpus to sample topics for each word token. VB consumes the longest training time due to its complicated digamma function calculation [76]. For a better illustration, we multiply VB's training time by 0.3. This result confirms that VB is time-consuming and unsuitable for big topic modeling.

The training time of BP is shorter than that of GS when $K$ is small, but is longer than that of GS when $K$ is large on ENRON, NYTIMES, and PUBMED data sets. The major reason is that BP requires $K$ iterations for the local message normalization,

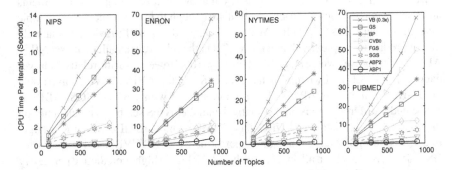

**Fig. 7.39** CPU time per iteration (seconds) as a function of topics $K \in \{100, 300, 500, 700, 900\}$ on NIPS, ENRON, NYTIMES, and PUBMED data sets. The notation $0.3x$ denotes the training time is multiplied by 0.3

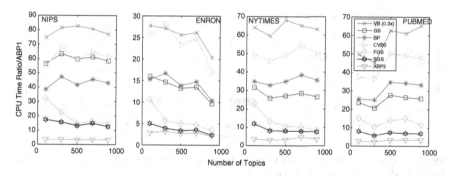

**Fig. 7.40** Ratio of CPU time per iteration over ABP1 as a function of topics $K \in \{100, 300, 500, 700, 900\}$ on NIPS, ENRON, NYTIMES, and PUBMED data sets. The notation $0.3x$ denotes the CPU time is multiplied by 0.3

while GS needs less than $K$ iterations because it stops visiting the rest topic space after sampling the target topic. Because CVB0 sweeps the entire word tokens instead of word indices at each iteration, it consumes approximately $N_d / W_d \approx 1.5$ training time of BP in Table 7.4. When $\lambda_d = \lambda_k = 0.1$, ABP1 needs only $1/100$ training time of BP. However, as ABP1 has to sort and update residuals, on average it consumes around $1/30$ training time of BP. When $\lambda_d = \lambda_k = 0.2$, ABP2 in anticipation requires four times longer training time than ABP1. But in practice ABP1 runs around three times faster than ABP2. Figure 7.39 demonstrates that at present ABP is the fastest batch LDA algorithm.

Figure 7.40 shows the ratio of CPU time per iteration over ABP1 as the benchmark. For a better illustration, we multiply the ratio of VB by 0.3. We see that ABP1 runs at least 100 times faster than VB on NIPS, NYTIMES, and PUBMED data sets. On ENRON set, we also see that ABP1 is close to 100 times faster than VB. The speedup effects are significant. For example, if VB uses three months for training LDA, ABP1 consumes only one day for the same task with much better generalization ability as shown in Fig. 7.38. ABP1 also runs around 20-40 times faster than both GS and BP. SGS is around 1.1-2 times faster than FGS, consistent with the result in [70].

When compared with the relatively faster FGS and SGS, ABP1 still runs around 10-20 times faster with the comparable predictive perplexity. ABP2 consistently consumes 2-3 times more training time than ABP1 because it sweeps more documents and topics. Both Figs. 7.38 and 7.40 reconfirm that ABP can be around 10-100 times faster than current state-of-the-art batch LDA algorithms with a comparable accuracy.

Technically, ABP can converge to a fixed-point as discussed in Sect. 7.4.2. Figure 7.41 shows the training perplexity as a function of number of training iterations when $K = 500$. We see that GS/FGS/SGS converges at the same number of iterations, and BP/CVB0 also converges at the same number of iterations with almost overlapped curves. Usually, GS/FGS/SGS uses 400–500 iterations, while BP/CVB0 needs 100–200 iterations to achieve convergence. Among all the algorithms, VB converges with the least number of iterations, which often uses around 80–150 iterations for convergence. Although VB sometimes achieves the lowest training perplexity

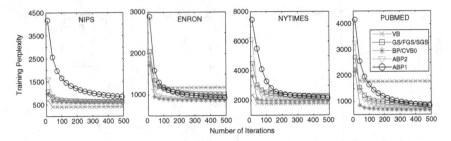

**Fig. 7.41** Training perplexity as a function of the number of iterations when $K = 500$ on NIPS, ENRON, NYTIMES, and PUBMED data sets

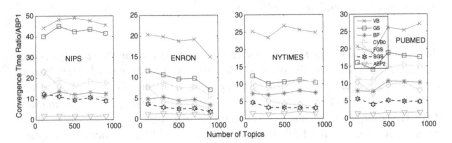

**Fig. 7.42** Convergence time ratio over ABP1 as a function of topics $K \in \{100, 300, 500, 700, 900\}$ on NIPS, ENRON, NYTIMES, and PUBMED data sets

on NIPS and NYTIMES data sets, it provides the highest predictive perplexity on unseen test set due to overfitting in Fig. 7.38. ABP2 converges with the more number of iterations than BP, which is caused by scanning the subset of corpus and searching the subset of topic space at each iteration. In our experiments, ABP2 uses around 300–500 iterations to achieve convergence. On NYTIMES and PUBMED data sets, ABP2 converges with almost the same number of iterations as GS/FGS/SGS. Furthermore, ABP1 converges with the largest number of training iterations to reach convergence on all data sets. In practice, ABP1 often uses 500–800 iterations for convergence, because at each iteration it has to scan 10 % corpus and search 10 % topics for the fast speed. However, even if we compare the training time until convergence, ABP1 is still the fastest algorithm. To confirm this, Fig. 7.42 summarizes the training time ratio until convergence over ABP1. We see that ABP1 still runs around 20–45 times faster than VB, 10–40 times faster than GS, and 5–20 times faster than CVB0, FGS/SGS, and BP when measured by the training time until convergence.

Figure 7.43 shows the top ten words of $K = 10$ topics extracted by VB (red), GS/FGS/SGS (blue), BP/CVB0 (black), and ABP (green) algorithms on the NIPS training set. For ABP, we choose $\lambda_d = \lambda_k = 0.5$ because $K = 10$ is small. We see that all algorithms can infer thematically meaningful topics, where most top ten words are consistent except the slightly different word ranking. There are two

| | |
|---|---|
| Topic 1 | network recognition system training neural word set speech input classifier<br>network recognition word system speech training set neural classifier model<br>**classifier recognition word set training classification speech model system class**<br>recognition training classifier set network word data classification speech system |
| Topic 2 | circuit neuron chip signal input network output system analog neural<br>signal circuit system chip neural analog output network input current<br>**circuit neural chip analog system input network output current voltage**<br>signal circuit chip analog system output neural input current sound |
| Topic 3 | network function learning neural weight algorithm input error result linear<br>network function learning weight neural input output unit error equation<br>**network unit input weight output training neural learning layer hidden**<br>network unit input weight output learning neural training layer hidden |
| Topic 4 | neuron model input cell network spike synaptic firing system pattern<br>neuron model cell input spike synaptic activity firing response pattern<br>**neuron model input cell synaptic spike network activity firing system**<br>neuron model cell input spike synaptic activity network firing response |
| Topic 5 | learning function algorithm action policy problem control optimal step reinforcement<br>learning algorithm function action problem result policy step theorem states<br>**learning control action system model task policy reinforcement step dynamic**<br>model control system learning movement controller robot motor position task |
| Topic 6 | network unit learning weight input output training hidden set neural<br>network unit learning input training weight output hidden task set<br>**function algorithm learning point problem linear number order result case**<br>function algorithm learning problem result action number bound policy set |
| Topic 7 | model data parameter function algorithm distribution gaussian set network vector<br>model data distribution parameter gaussian algorithm probability component method density<br>**model data distribution probability parameter gaussian method mean noise density**<br>model data distribution parameter method gaussian function set algorithm probability |
| Topic 8 | image visual model field motion cell direction object map images<br>image visual field object images motion map direction feature cell<br>**visual model field motion cell image direction object map eye**<br>image visual field motion cell images object map model direction |
| Topic 9 | model control system network learning neural movement input motor position<br>model control system learning movement robot controller motor position eye<br>**signal frequency component filter sound system auditory information analysis data**<br>set model algorithm structure problem representation rules graph vector cluster |
| Topic 10 | data model set error training algorithm network function method learning<br>data set training error function algorithm vector method problem network<br>**vector data feature set image problem features point representation algorithm**<br>function learning point algorithm network vector linear matrix equation order |

**Fig. 7.43** Top ten words of ten topics on NIPS: VB (*red*), GS/FGS/SGS (*blue*), BP/CVB0 (*black*), and ABP (*green*)

subjective measures for the interpretability of discovered topics: word intrusion and topic intrusion [14]. The former is the number of word intruders in each topic, while the latter is the number of topic intruders in each document. According to our experience, the word and topic intrusions are comparable among all the algorithms. So, given the similar interpretable topics, we advocate ABP for fast topic modeling.

Although at each iteration ABP updates and normalizes only the subset of $K$-tuple messages (7.37), it devotes almost the equal number of iterations to different topics $K$ with two main reasons. First, at each iteration, different documents contribute different topics according to residuals (7.35). So, each topic will be visited by almost the equal number of times due to the LDA's clustering property [76] across all documents. Second, ABP uses the dynamical scheduling of residuals shown in Fig. 7.11 at each iteration, which ensures that all $K$-tuple messages will be updated. Figure 7.43 demonstrates that most topics extracted by ABP have almost the same top ten words as VB, GS, and BP even if ABP scans 50 % documents and 50 % topics at each iteration.

### 7.4.3.4  Comparison with Online and Parallel Algorithms

When $K = 500$, $\alpha = 2/K$, and $\beta = 0.01$, we compare ABP1 with two state-of-the-art online LDA algorithms including OGS [70],[12] OVB [28],[13] and RVB [61], where we use their default parameters. The mini-batch size is 16 for the NIPS set, and 1024 for other data sets. Based on the same test set in Table 7.4, the algorithms OGS, OVB, and RVB use the remaining 292,500, and 8,160,000 documents from the original NYTIMES and PUBMED data sets [48] for online inference.

Figure 7.44 shows the predictive perplexity as a function of training time in seconds (ln-scale) on four data sets. Clearly, ABP1 achieves a lower predictive perplexity by using only a fraction of training time consumed by OGS, OVB, and RVB. There are three reasons for this significant speedup. First, OGS is derived from SGS for each mini-batch, which converges slower than ABP1 in Fig. 7.41. Second, the computationally expensive digamma functions significantly slow down both OVB and RVB. Finally, OGS and OVB require loading each mini-batch from the hard disk into memory, consuming more time than in-memory bath algorithms when the number of mini-batch is large. RVB is slightly slower than OVB because of additional scheduling costs, but it often achieves a relatively lower perplexity when compared with OVB. According to [42], SOI is around twice faster than OVB in practice, so ABP1 is still much faster than SOI to reach a lower predictive perplexity when compared with the same benchmark OVB.

Figure 7.44 also compares batch VB, BP, and SGS with OGS, OVB, and RVB. We see that OGS, OVB, and RVB perform worse than VB, BP, and SGS on the NIPS and ENRON data sets, partly because the online gradient descents introduce noises [10] on the relatively smaller data sets. However, after OGS, OVB, and RVB scan more

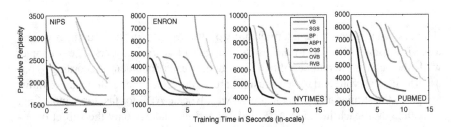

**Fig. 7.44**  Predictive perplexity as a function of training time in seconds (ln-scale) on NIPS, ENRON, NYTIMES, and PUBMED data sets when $K = 100$. The notation $0.3x$ denotes the predictive perplexity is multiplied by 0.3

---

[12] http://mallet.cs.umass.edu/.

[13] http://www.cs.princeton.edu/~blei/topicmodeling.html.

**Table 7.5** Average training time (s) per iteration

| Data sets | PGS (32 processors) | ABP1 |
|-----------|---------------------|------|
| NIPS | 0.38 | 0.09 |
| ENRON | 1.69 | 1.39 |
| NYTIMES | 0.68 | 0.51 |
| PUBMED | 0.69 | 0.55 |

training documents on the NYTIMES and PUBMED data sets, they can achieve comparable or even lower perplexity values than SGS and VB, respectively. These results are consistent with [28, 70]. Overall, BP converges to the lowest predictive perplexity on four data sets consistent with those shown in Fig. 7.38. With more and more training data streams, online algorithms may arrive at lower perplexity values than batch counterparts with significantly less memory consumptions. However, since memory becomes cheaper and cheaper, RAMCloud is available and in-memory computation of batch algorithms for big data may be possible in near future [47].

When $K = 500, \alpha = 2/K, \beta = 0.01$, and $T = 500$, we also compare ABP1 with one of the state-of-the-art parallel LDA algorithms such as PGS [45] using 32 processors. PGS is the parallel implementation of an approximate inference algorithm of GS but obtains almost the same predictive perplexity values as GS in Fig. 7.38, which indicates that the topic modeling accuracy of PGS is still lower than ABP1 except for the NIPS set. As it would be expected, PGS on 32 processors should achieve 1/32 training time of GS. However, PGS cannot achieve the maximum parallel efficiency due to the extensive communication and synchronization delay. After the 32 processors have finished scanning the distributed data sets, they need to communicate the multinomial parameter matrix $\phi_{K \times W}$ (7.11) for synchronization. This matrix $\phi$ increases with the number of topics $K$, adding the burden of communication and synchronization for the large number of topics $K$.

Table 7.5 shows the average training time per iteration between ABP1 and PGS. Obviously, ABP1 is still much faster than PGS. One major reason for such a speedup is that PGS searches 500 topics at each iteration, while ABP1 selects only 50 topics at each iteration. Indeed, the large number of topics can be also handled by the parallel FGS (PFGS) [48] and SGS (PSGS) implementations. Although PGS can further accelerate the topic modeling speed by adding more processors, the parallel efficiency will be further reduced due to more communication costs among more processors. In addition, the price-to-performance ratio of PGS is much higher than ABP1. Overall, ABP is a promising tool for fast topic modeling.

## 7.5 Human Action Recognition

We examine our algorithms in Figs. 7.18 and 7.20 on two publicly available data sets that have been widely used in action recognition: the KTH [52] human motion data set and the Weizmann [7] human action data set. Although the action recognition performance on these benchmark data sets seems saturating, i.e., many state-of-the-art methods can achieve around 80–90 % recognition accuracy, we show that our algorithms can achieve a comparable recognition results and, more importantly, that our extended T2 FTMs significantly outperform two topic models: the baseline LDA [46] and L-LDA [62].

### 7.5.1 Feature Extraction and Vocabulary Formation

We represent each video sequence as a bag of visual words by extracting the spatial–temporal interest points, which has been confirmed as a describable and robust representation of human actions in previous works [19, 46]. Among the available interest point detectors for video data, we choose to use the separable linear filter method proposed in [46] because it can produce sufficient large number of interest points. By linear filters, any region with spatially distinguishing characteristics undergoing a periodic motion can induce a strong response; however, areas undergoing pure translational motion, or without spatially distinguishing features will not induce a strong response. The detailed response function of separable linear filters can be referred to [46].

Figure 7.13a shows some examples of spatial–temporal interest points detected using the method of separable linear filters in [46]. Each white box corresponds to a spatial–temporal cuboid centered at the detected interest point. The cuboid size is approximately six times the scales along 3 dimensions $\{x, y, t\}$, where $\{x, y\}$ are the spatial dimensions and $\{t\}$ is the temporal dimension. The scales are one spatial and one temporal parameters $\{\sigma, \tau\}$ in the response function of separate linear filters [19, 46]. For each cuboid, we calculate the brightness gradients on three directions $\{x, y, t\}$ as a descriptor for each spatial–temporal cuboid. The computed gradients are concatenated to form a feature vector for each cuboid, whose size is the number of gradients times three directions. Because the dimension of each feature vector is very high, we use the principal component analysis (PCA) [46] to project the high-dimensional feature vector to a lower dimension of 100. Finally, we construct a vocabulary list of all cuboids referred to as visual words by clustering their feature vectors using the $k$-means algorithm with Euclidean distance as the clustering metric. The center of each resulting cluster is defined to be a spatial–temporal vocabulary word. The vocabulary size $W$ is equal to the number of clusters $K$ in $k$-means algorithm provided by users. The larger vocabulary size often yields better recognition results. In test videos, newly detected visual words (cuboid features) are mapped to this vocabulary list for a unique cluster membership. Therefore, a video sequence can be represented as a collection of spatial–temporal words from the vocabulary in Fig. 7.13b. More details on spatial–temporal motion word extraction can be found in [19, 46]. It is worth noting that another work [62] proposes a global optical flow feature for each frame in a video sequence as a motion word, which shows a slightly better recognition performance than local cuboid features. The global features, however, require an automatic preprocessing step to track and stabilize the video sequences, and also require the substraction of backgrounds. So, the quality of global features depends highly on tacking and stabilization results at preprocessing steps. Since different visual word schemes, either local or global features, can be used in our algorithms, we focus on comparing different models on the same local cuboid features [46] for a fair comparison in the next subsections. We believe that a model having a good performance on local features will also have a good performance on global features without much tuning.

### 7.5.2  Results on KTH Data Set

The KTH human motion data set [52] contains six types of human actions (walking, jogging, running, boxing, hand waving, and hand clapping) performed around 4 times by 25 persons in four different scenarios: outdoors, outdoors with scale variation, outdoors with different clothes, and indoors. Figure 7.13a shows some sample frames of each action. KTH has a total of 598 video sequences. Each video contains one action. Since the data set is very large and all the motions are periodic, we use a subset of the data set which includes each person repeating each activity 8 times for about 4 s with a total of $598 \times 2 = 1196$ clips. Half of the sequences is used to build the vocabulary of visual words, and the other half is used for training and testing purposes. We use the same parameters $\{\sigma = 2, \tau = 2.5\}$ in response function of separate linear filters to extract spatial–temporal interest points as [46]. On average, there are 50 motion words per video. To test the efficiency of our approach for the categorization task, we divide the human action sequences into 25 groups, one per person. We adopt the leave-one-out testing paradigm that we learn a model from the videos of 24 persons and test the videos of the remaining persons. The recognition accuracy is the average of 25 runs.

We train the IT2 FTM, VT2 FTM, LDA [46], and L-LDA [62] models for recognizing human action categories. For a fair comparison, LDA and L-LDA are learned by the BP algorithm [76, 86], which has been confirmed more accurate and faster than existing inference algorithms in topic modeling applications. We set the same hyperparameters $\alpha = \beta = 0.01$ for all experiments. LDA is unsupervised method and usually requires $T = 300$ iterations for convergence. L-LDA and T2 FTMs are supervised method and only $T = 50$ iterations is enough for convergence. Our T2 FTM algorithms are very efficient. Most of the computation is spent on the feature extraction and vocabulary formation. After the BOW representation for each video sequence is available, learning IT2 and VT2 FTMs usually takes a few minutes, and inference on a test video takes a few seconds because the learning algorithms in Figs. 7.18 and 7.20 converge very fast. When compared with L-LDA, our algorithms are slightly slower because IT2 FTM additionally estimates the standard deviation and does interval normalization during message passing, while VT2 FTM additionally updates the message sparseness at each iteration. For IT2 FTM, we test different parameters $u \in \{0.5, 1, 2, 3\}$ for message uncertainty in Eqs. (7.51)–(7.54). We find that IT2 FTM achieves the best performance when $u = 2$ in our experiments.

**Fig. 7.45** Classification accuracy of IT2 FTM, VT2 FTM, LDA, and L-LDA versus the vocabulary size

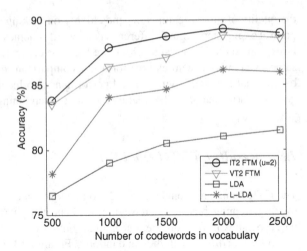

Because the performance of different $u$ is relatively stable ($\pm 3\%$), we show only the classification performance when $u = 2$ for simplicity.

Generally, the size of the vocabulary is related to the classification performance [46, 62]. So, we examine the vocabulary size $W$ as shown in Fig. 7.45. It is obvious that the larger size of vocabulary leads to a higher accuracy for all models. The performance seems saturated, however, when the vocabulary size is greater than 2,000. We see that IT2 FTM ($u = 2$) is slightly better than VT2 FTM. But, VT2 FTM has some advantages because IT2 FTM requires manually tuning additional free parameter $u$ in different applications. Our hypothesis for this outcome is that IT2 FTM uses the standard deviation to account for message uncertainties, which is better than VT2 FTM using the sparseness-based method to describe message uncertainties. Overall, both T2 FTMs outperform the baseline LDA and L-LDA because these baseline models do not consider higher-order uncertainties of motion words. It is reasonable that LDA performs the worst because it is an unsupervised method without incorporating class labels during training.

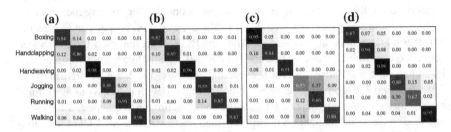

**Fig. 7.46** Confusion matrix for per-video classification of **a** IT2 FTM (overall accuracy 89.33 %) and **b** VT2 FTM (overall accuracy 88.83 %) **c** LDA (overall accuracy 81.05 %) and **d** L-LDA (overall accuracy 86.17 %) using 2,000 codewords

**Table 7.6** Comparison of different methods using the KTH data set ($W = 2000$)

| Methods | Accuracy (%) |
|---|---|
| IT2 FTM ($u = 2$) | 89.33 |
| VT2 FTM | 88.83 |
| L-LDA [50, 62] | 86.17 |
| LDA [46] | 81.50 |
| SVM [19] | 81.17 |
| SVM [52] | 71.72 |

The confusion matrix for per-video classification using 2,000 codewords is shown in Fig. 7.46. Each row in the confusion matrix corresponds to the ground truth category, and each column corresponds to the predicted action class. We see that actions "boxing" and "clapping" are often misclassified each other, while actions "running" and "jogging" are often misclassified each other. These results are reasonable because these actions are very close. When compared with LDA and L-LDA, we find that our T2 FTMs improve the recognition accuracy of "running" and "jogging" significantly. For example, the accuracy of IT2 FTM is 27 % and 20 % higher than that of LDA and L-LDA, respectively. We may use the "fuzzy classification boundary" [82] to explain this outcome. Misclassification often occurs in data points close to the classification boundary. T2 FS-based classifiers changes the "crisp" classification boundary to the "fuzzy" one so that it may account for more uncertainties of data points close to the "fuzzy" boundary.

Table 7.6 compares the per-video recognition accuracy (the correctly predicted videos/the total number of videos) among different state-of-the-art methods when $W = 2000$. We see that IT2 FTM outperforms L-LDA by around 3 %, which confirms the effectiveness of using T2 FS for handling uncertainty of motion word. We also compare our results with the discriminative model, support vector machines (SVMs) [19], with the same experimental settings. Similar to previous works [46, 62], we find that SVMs perform worse than generative models like LDA. Our extended IT2 FTM ($u = 2$) improves the accuracy by at least 10 % when compared with SVM [19]. We accept the fact that the comparison to the original L-LDA [62] is not complete, since L-LDA is trained on the global features rather than the local features in our experimental settings. However, our T2 FTMs provide new solutions to address motion word uncertainties, which have been largely ignored by both LDA and L-LDA models.

### 7.5.2.1 Results on Weizmann Dataset

The Weizmann human action data set [7] contains 90 video sequences with 10 action categories performed by 9 persons. Similar to [46], we set the parameters of response function in separate linear filters as $\{\sigma = 1.2, \tau = 1.2\}$, and also reduce the dimensionality of feature vectors by PCA to 100 dimensions. Due to the smaller size of data set, we use all video sequences to build the vocabulary. Again, we use the

**Fig. 7.47** Ten categories of motions in the Weizmann human action data set

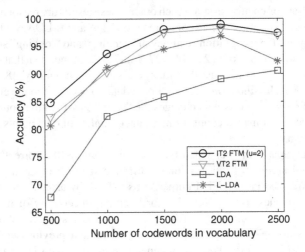

**Fig. 7.48** Classification accuracy of IT2 FTM, VT2 FTM, LDA, and L-LDA versus the vocabulary size

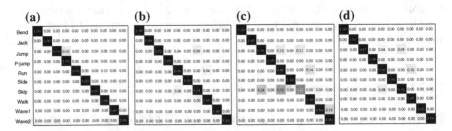

**Fig. 7.49** Confusion matrix for per-video classification of **a** IT2 FTM (overall accuracy 99.00 %) and **b** VT2 FTM (overall accuracy 98.20 %) **c** LDA (overall accuracy 89.10 %) and **d** L-LDA (overall accuracy 96.90 %) using 2000 codewords

**Table 7.7** Comparison of different methods using the Weizmann dataset ($W = 2000$)

| Methods | Accuracy (%) |
|---|---|
| IT2 FTM ($u = 2$) | 99.00 |
| VT2 FTM | 98.20 |
| L-LDA [50, 62] | 96.90 |
| LDA [46] | 89.10 |

leave-one-out scheme to examine the effectiveness of our approach in classification. For each run, we train a model from the videos of eight persons, and test those of remaining persons. The recognition accuracy is calculated by the average of nine runs. Figure 7.47 shows ten typical motions in the Weizmann human action data set.

Figure 7.48 shows the classification accuracy versus the vocabulary size $W \in \{500, 1000, 1500, 2000, 2500\}$. Similar to Fig. 7.45, we find that the classification rate increases with the size of vocabulary but has saturated when $W > 2000$. Figure 7.49 shows the confusion matrices for 10 classes produced by IT2 FTM ($u = 2$), VT2 FTM, LDA, and L-LDA models. We see that LDA and L-LDA are often confused by similar action classes, such as "skip" with "jump" and "run", and "run" with "walk". In contrast, our T2 FTMs significantly improve the recognition accuracy in these three confusing actions "skip", "jump", and "run". For example, VT2 FTM improves LDA and L-LDA by around 36 % and 4.8 % in terms of the accuracy rate, respectively. Table 7.7 summarizes the average recognition accuracy of these methods. We see that IT2 FTM is slightly better than VT2 FTM, confirming our hypothesis that the standard deviation of word messages is a good way to measure uncertainties of motion words.

# References

1. AlSumait, L., Barbará, D., Domeniconi, C.: On-line LDA: adaptive topic models for mining text streams with applications to topic detection and tracking. In: ICDM, pp. 3–12 (2008)
2. Asuncion, A., Welling, M., Smyth, P., Teh, Y.W.: On smoothing and inference for topic models. In: UAI, pp. 27–34 (2009)
3. Asuncion, A.: Approximate Mean Field for Dirichlet-Based Models. In: ICML Workshop on Topic Models (2010)
4. Banerjee, A., Basu, S.: Topic models over text streams: a study of batch and online unsupervised learning. In: SIAM Data Mining, pp. 437–442 (2007)
5. Berge, C.: Hypergraphs. North-Holland, Amsterdam (1989)
6. Bishop, C.M.: Pattern Recognition and Machine Learning. Springer, New York (2006)
7. Blank, M., Gorelick, L., Shechtman, E., Irani, M., Basri, R.: Actions as space-time shapes. In: ICCV, pp. 1395–1402 (2005)
8. Blei, D.M., Ng, A.Y., Jordan, M.I.: Latent Dirichlet allocation. J. Mach. Learn. Res. **3**, 993–1022 (2003)
9. Blei, D.M.: Introduction to probabilistic topic models. Commun. ACM **55**, 77–84 (2012)
10. Bottou, L., Cun, Y.L.: Large scale online learning. In: NIPS, pp. 217–224 (2004)
11. Bottou, L.: Online Learning and Stochastic Approximations. Cambridge University Press, Cambridge (1998)

12. Buntine, W.L.: Variational extensions to EM and multinomial PCA. In: ECML, pp. 23–34 (2002)
13. Canini, K.R., Shi, L., Griffths, T.L.: Online inference of topics with latent Dirichlet allocation. In: AISTATS, pp. 65–72 (2009)
14. Chang, J., Boyd-Graber, J., Gerris, S., Wang, C., Blei, D.: Reading tea leaves: How humans interpret topic models. In: NIPS, pp. 288–296 (2009)
15. Chang, C.C., Lin, C.J.: LIBSVM: a library for support vector machines. ACM Trans. Intell. Syst. Technol. 2, 27:1–27:27 (2011)
16. Chang, J., Blei, D.M.: Hierarchical relational models for document networks. Ann. Appl. Stat. 4(1), 124–150 (2010)
17. Cutler, R., Davis, L.S.: Robust real-time periodic motion detection, analysis, and applications. IEEE Trans. Pattern Anal. Mach. Intell. 22, 781–796 (2000)
18. Dempster, A.P., Laird, N.M., Rubin, D.B.: Maximum likelihood from incomplete data via the EM algorithm. J. Roy. Stat. Soc. B 39, 1–38 (1977)
19. Dollár, P., Rabaud, V., Cottrell, G., Belongie, S.: Behavior recognition via sparse spatio-temporal features. Vis. Surveill. Perform. Eval. Tracking Surveill 2, 65–72 (2005)
20. Eisenstein, J., Xing, E.: The CMU 2008 political blog corpus. Technical report, Carnegie Mellon University (2010)
21. Elidan, G., McGraw, I., Koller, D.: Residual belief propagation: informed scheduling for asynchronous message passing. In: UAI, pp. 165–173 (2006)
22. Fei-Fei, L., Perona, P.: A Bayesian hierarchical model for learning natural scene categories. In: CVPR, pp. 524–531 (2005)
23. Foundation, A.S., et al.: Apache Mahout (2010). http://mloss.org/software/view/144/
24. Frey, B.J., Dueck, D.: Clustering by passing messages between data points. Science 315(5814), 972–976 (2007)
25. Globerson, A., Chechik, G., Pereira, F., Tishby, N.: Euclidean embedding of co-occurrence data. J. Mach. Learn. Res. 8, 2265–2295 (2007)
26. Griffiths, T.L., Steyvers, M.: Finding scientific topics. Proc. Natl. Acad. Sci. 101, 5228–5235 (2004)
27. Heinrich, G.: Parameter estimation for text analysis. Technical report, University of Leipzig (2008)
28. Hoffman, M., Blei, D., Bach, F.: Online learning for latent Dirichlet allocation. In: NIPS, pp. 856–864 (2010)
29. Hofmann, T.: Unsupervised learning by probabilistic latent semantic analysis. Mach. Learn. 42, 177–196 (2001)
30. Hoyer, P.O.: Non-negative matrix factorization with sparseness constraints. J. Mach. Learn. Res. 5, 1457–1469 (2004)
31. Karnik, N.N., Mendel, J.M.: Centroid of a type-2 fuzzy set. Inf. Sci. 132, 195–220 (2001)
32. Karnik, N.N., Mendel, J.M.: Operations on type-2 fuzzy sets. Fuzzy Sets Syst. 122, 327–348 (2001)
33. Kschischang, F.R., Frey, B.J., Loeliger, H.A.: Factor graphs and the sum-product algorithm. IEEE Trans. Inf. Theory 47(2), 498–519 (2001)
34. Lee, D.D., Seung, H.S.: Learning the parts of objects by non-negative matrix factorization. Nature 401, 788–791 (1999)
35. Liang, Q., Mendel, J.M.: MPEG VBR video traffic modeling and classification using fuzzy technique. IEEE Trans. Fuzzy Syst. 9(1), 183–193 (2001)
36. Liu, Z., Zhang, Y., Chang, E., Sun, M.: Plda+: parallel latent Dirichlet allocation with data placement and pipeline processing. ACM Trans. Intell. Syst. Technol. 2(3), 26 (2011)
37. McCallum, A.K., Nigam, K., Rennie, J., Seymore, K.: Automating the construction of internet portals with machine learning. Inf. Retrieval 3(2), 127–163 (2000)
38. Mendel, J.M., John, R.I.B., Liu, F.: Interval type-2 fuzzy logic systems made simple. IEEE Trans. Fuzzy Syst. p. accepted for publication (2005)
39. Mendel, J.M.: Uncertainty, fuzzy logic, and signal processing. Signal Process. 80(6), 913–933 (2000)

40. Mendel, J.M.: Uncertain Rule-based Fuzzy Logic Systems: Introduction and New Directions. Prentice-Hall, Upper Saddle River (2001)
41. Mendel, J.M., John, R.I.B.: Type-2 fuzzy sets made simple. IEEE Trans. Fuzzy Syst. **10**(2), 117–127 (2002)
42. Mimno, D., Hoffman, M.D., Blei, D.M.: Sparse stochastic inference for latent Dirichlet allocation. In: ICML (2012)
43. Minka, T., Lafferty, J.: Expectation-propagation for the generative aspect model. In: UAI, pp. 352–359 (2002)
44. Mitchell, H.: Pattern recognition using type-II fuzzy sets. Inf Sci **170**, 409–418 (2005)
45. Newman, D., Asuncion, A., Smyth, P., Welling, M.: Distributed algorithms for topic models. J. Mach. Learn. Res. **10**, 1801–1828 (2009)
46. Niebles, J., Wang, H., Fei-Fei, L.: Unsupervised learning of human action categories using spatial-temporal words. Int. J. Comput. Vis. **79**, 299–318 (2008)
47. Ousterhout, J., et al.: The case for RAMClouds: scalable high-performance storage entirely in DRAM. SIGOPS Oper. Syst. Rev. **43**(4), 92–105 (2012)
48. Porteous, I., Newman, D., Ihler, A., Asuncion, A., Smyth, P., Welling, M.: Fast collapsed Gibbs sampling for latent Dirichlet allocation. In: KDD, pp. 569–577 (2008)
49. Pruteanu-Malinici, I., Ren, L., Paisley, J., Wang, E., Carin, L.: Hierarchical Bayesian modeling of topics in time-stamped documents. IEEE Trans. Pattern Anal. Mach. Intell. **32**(6), 996–1011 (2010)
50. Ramage, D., Hall, D., Nallapati, R., Manning, C.D.: Labeled LDA: A supervised topic model for credit attribution in multi-labeled corpora. In: Empirical Methods in Natural Language Processing, pp. 248–256 (2009)
51. Rosen-Zvi, M., Griffiths, T., Steyvers, M., Smyth, P.: The author-topic model for authors and documents. In: UAI, pp. 487–494 (2004)
52. Schuldt, C., Laptev, I., Caputo, B.: Recognizing human actions: a local SVM approach. IN: ICPR, vol. 3, 32–36 (2004)
53. Sevastjanov, P., Bartosiewicz, P., Tkacz, K.: A new method for normalization of interval weights. Parallel Process. Appl. Math. **6068**, 466–474 (2010)
54. Shi, J., Malik, J.: Normalized cuts and image segmentation. IEEE Trans. Pattern Anal. Machine Intell. **22**(8), 888–905 (2000)
55. Smola, A., Narayanamurthy, S.: An architecture for parallel topic models. In: PVLDB, pp. 703–710 (2010)
56. Sullivan, J., Carlsson, S.: Recognizing and tracking human actions. In: ECCV, pp. 629–644 (2002)
57. Tappen, M.F., Freeman, W.T.: Comparison of graph cuts with belief propagation for stereo, using identical MRF parameters. In: ICCV, pp. 900–907 (2003)
58. Teh, Y.W., Newman, D., Welling, M.: A collapsed variational Bayesian inference algorithm for latent Dirichlet allocation. In: NIPS, pp. 1353–1360 (2007)
59. Thiesson, B., Meek, C., Heckerman, D.: Accelerating EM for large databases. Mach. Learn. **45**, 279–299 (2001)
60. Wahabzada, M., Kersting, K., Pilz, A., Bauckhage, C.: More influence means less work: fast latent Dirichlet allocation by influence scheduling. In: CIKM, pp. 2273–2276 (2011)
61. Wahabzada, M., Kersting, K.: Larger residuals, less work: active document scheduling for latent Dirichlet allocation. In: ECML/PKDD, pp. 475–490 (2011)
62. Wang, Y., Mori, G.: Human action recognition by semilatent topic models. IEEE Trans. Pattern Anal. Mach. Intell. **31**(10), 1762–1774 (2009)
63. Wang, X.G., Ma, X.X., Grimson, W.E.L.: Unsupervised activity perception in crowded and complicated scenes using hierarchical Bayesian models. IEEE Trans. Pattern Anal. Mach. Intell. **31**(3), 539–555 (2009)
64. Welling, M., Rosen-Zvi, M., Hinton, G.: Exponential family harmoniums with an application to information retrieval. In: NIPS, pp. 1481–1488 (2004)
65. Winn, J., Bishop, C.M.: Variational message passing. J. Mach. Learn. Res. **6**, 661–694 (2005)

66. Wu, D., Nie, M.: Comparison and practical implementation of type-reduction algorithms for type-2 fuzzy sets and systems. In: FUZZ-IEEE, pp. 2131–2138 (2011)
67. Wu, H., Mendel, J.M.: Classification of battlefield ground vehicles using acoustic features and fuzzy logic rule-based classifiers. IEEE Trans. Fuzzy Syst. **15**(1), 56–72 (2007)
68. Yan, J., Liu, Z.Q., Gao, Y., Zeng, J.: Communication-efficient parallel belief propagation for latent Dirichlet allocation (2012). arXiv:1206.2190v1 [cs.LG]
69. Yan, F., Xu, N., Qi, Y.: Parallel inference for latent Dirichlet allocation on graphics processing units. In: NIPS, pp. 2134–2142 (2009)
70. Yao, L., Mimno, D., McCallum, A.: Efficient methods for topic model inference on streaming document collections. In: KDD, pp. 937–946 (2009)
71. Yu, H.F., Hsieh, C.J., Chang, K.W., Lin, C.J.: Large linear classification when data cannot fit in memory. In: KDD, pp. 833–842 (2010)
72. Zeng, J., Cao, X.Q., Liu, Z.Q.: Memory-efficient topic modeling (2012). arXiv:1206.1147 [cs.LG]
73. Zeng, J., Cao, X.Q., Liu, Z.Q.: Residual belief propagation for topic modeling (2012). arXiv:1201.0838v2 [cs.LG]
74. Zeng, J., Cheung, W.K., Li, C.H., Liu, J.: Coauthor network topic models with application to expert finding. In: IEEE/WIC/ACM WI-IAT, pp. 366–373 (2010)
75. Zeng, J., Cheung, W.K.W., Li, C.H., Liu, J.: Multirelational topic models. In: ICDM, pp. 1070–1075 (2009)
76. Zeng, J., Cheung, W.K., Liu, J.: Learning topic models by belief propagation. IEEE Trans. Pattern Anal. Mach. Intell. (2012). arXiv:1109.3437v4 [cs.LG]
77. Zeng, J., Liu, Z.Q., Cao, X.Q.: Online belief propagation for topic modeling (2012). arXiv:1206.1147 [cs.LG]
78. Zeng, J., Liu, Z.Q.: Interval type-2 fuzzy hidden Markov models. In: Proceedings of FUZZ-IEEE pp. 1123–1128 (2004)
79. Zeng, J., Liu, Z.Q.: Type-2 fuzzy hidden Markov models to phoneme recognition. In: 17th International Conference on Pattern Recognition 1, 192–195 (2004)
80. Zeng, J., Liu, Z.Q.: Type-2 fuzzy Markov random fields and their application to handwritten Chinese character recognition. IEEE Trans. Fuzzy Syst. **16**(3), 747-760 (2008)
81. Zeng, J., Liu, Z.Q.: Type-2 fuzzy Markov random fields to handwritten character recognition. In: 18th International Conference on Pattern Recognition pp. 1162–1165 (2006)
82. Zeng, J., Liu, Z.Q.: Type-2 fuzzy sets for handling uncertainty in pattern recognition. In: Proceedings of FUZZ-IEEE pp. 6597–6602 (2006)
83. Zeng, J., Liu, Z.Q.: Type-2 fuzzy hidden Markov models and their application to speech recognition. IEEE Trans. Fuzzy Syst. **14**(3), 454–467 (2006)
84. Zeng, J., Xie, L., Liu, Z.Q.: Type-2 fuzzy Gaussian mixture models. Pattern Recognit **41**(12), 3636–3643 (2008)
85. Zeng, J., Liu, Z.Q.: Type-2 fuzzy Markov random fields and their application to handwritten Chinese character recognition. IEEE Trans. Fuzzy Syst. **16**(3), 747–760 (2008)
86. Zeng, J.: A topic modeling toolbox using belief propagation. J. Mach. Learn. Res. **13**, 2233–2236 (2012)
87. Zhai, K., Boyd-Graber, J., Asadi, N., Alkhouja, M.: Mr. LDA: a flexible large scale topic modeling package using variational inference in mapreduce. In: WWW (2012)
88. Zhu, S., Zeng, J., Mamitsuka, H.: Enhancing MEDLINE document clustering by incorporating MeSH semantic similarity. Bioinformatics **25**(15), 1944–1951 (2009)
89. Zhu, S., Takigawa, I., Zeng, J., Mamitsuka, H.: Field independent probabilistic model for clustering multi-field documents. Inf. Process. Manage. **45**(5), 555–570 (2009)
90. Zipf, G.K.: Human behavior and the principle of least effort. Addison-Wesley, Cambridge (1949)

# Chapter 8
# Conclusions and Future Work

**Abstract** This chapter summarizes the book and envisions future works.

## 8.1 Conclusions

Chapter 1 is an overview of this book.

- Pattern recognition is concerned with three central issues: (1) feature extraction, (2) hypothesis selection, and (3) learning algorithm. All issues are involved with prior knowledge about specific applications.
- This book mainly deals with two types of uncertainties in pattern recognition, i.e., randomness and fuzziness.
- Graphical models as hypotheses can (1) statistical-structurally represent patterns, and (2) have efficient learning and decoding algorithms.
- Type-2 fuzzy sets can describe bounded uncertainty in both feature and hypothesis spaces, and handle randomness and fuzziness simultaneously.
- The main contribution of this book lies in four aspects: (1) We introduce graphical models for the statistical-structural pattern recognition paradigm; (2) We use type-2 fuzzy sets for handling both randomness and fuzziness uncertainties in pattern recognition; (3) We extend classical graphical models such as Gaussian mixture models, hidden Markov models, Markov random fields and latent Dirichlet allocation to type-2 fuzzy graphical models for handling more uncertainties; and (4) We apply the proposed approaches to real-world pattern classification problems such as speech, handwriting, and human action recognition.

In Chap. 2, we formulate pattern recognition as a labeling problem. Many pattern recognition problems can be posed as labeling problems to which the solution is a set of linguistic labels assigned to extracted features from speech signals, image pixels, and image regions. Graphical models use Markov properties to measure a local probability on the labels within the neighborhood system. The Bayesian decision theory guarantees the best labeling configuration according to the *maximum a posteriori* criterion.

© Tsinghua University Press, Beijing and Springer-Verlag Berlin Heidelberg 2015
J. Zeng and Z.-Q. Liu, *Type-2 Fuzzy Graphical Models for Pattern Recognition*,
Studies in Computational Intelligence 591, DOI 10.1007/978-3-662-44690-4_8

Chapter 3 introduces type-2 fuzzy sets and type-2 fuzzy logic systems. Type-2 fuzzy sets have two new concepts: secondary membership function and footprint of uncertainty. The secondary membership function evaluates the primary memberships, and the footprint of uncertainty describes bounded uncertainty of the primary memberships. In wavy slice representation, the type-2 fuzzy set can be viewed as embedded with many type-1 fuzzy sets. General type-2 fuzzy sets' operations are prohibitive, whereas interval type-2 fuzzy sets' operations are concerned with simple interval arithmetic. In type-2 fuzzy logic systems, nonsingleton fuzzifier can map crisp input to fuzzy numbers, and the output of the system is a type-2 fuzzy set so that type-reduction and defuzzification are needed. The Karnic-Mendel algorithm is special for the type-reduction of interval type-2 fuzzy set. Using one type-2 fuzzy set for uncertain feature space, and the other type-2 fuzzy set for uncertain hypothesis space, pattern recognition can be performed by a similarity measure of these two sets. On the other hand, if the input feature is nonsingleton fuzzified, and the rule base is modeled by type-2 fuzzy sets, pattern recognition is dependent on the type-reduced set of the type-2 fuzzy logic system. Because of fuzziness, the classical Bayesian decision theory is no longer the best decision rule for classification. Thus, we extend it by type-2 fuzzy set operations referred to as the type-2 fuzzy Bayesian decision theory, which provides a general rule to make decision when both randomness and fuzziness exist in pattern recognition.

In Chap. 4, we introduce type-2 fuzzy Gaussian mixture models (T2 GMMs). In real-world applications, we often encounter uncertain feature space, which results in a mismatch between the GMM-based class-conditional densities and the underlying distributions. To reflect such uncertainty as mismatch, we use type-2 fuzzy sets to describe uncertain parameters in Gaussian mixture models. By information theory, we explain why the lower and upper boundaries of the T2 FGMM can provide additional information for classifying outliers. Multivariate Gaussian primary membership function with uncertain mean vector and covariance matrix is the most widely used type-2 fuzzy membership functions in this book. In the proposed classification system, we first build a T2 FGMM for each category, and evaluate each training sample by all models. As a result, we obtain a feature vector composed of interval likelihoods. In this case, the maximum likelihood criterion is not suitable for decision-making, so we adopt the generalized linear model to fulfill this task automatically.

Chapter 5 studies type-2 fuzzy hidden Markov models (T2 FHMMs) and their applications to speech recognition. The type-2 fuzzy HMM is an extension of the HMM. The corresponding type-2 fuzzy forward–backward, Viterbi, and Baum–Welch algorithms have been developed using type-2 fuzzy sets operations. Especially, the interval type-2 fuzzy HMM has practical use. The IT2 FHMM can effectively handle babble noise and dialect uncertainties in speech data besides a better classification performance than the classical HMM. We have three methods to make decisions based on the output interval sets from the IT2 FHMM: (1) Rank the output interval sets directly, (2) Use the centroid of interval sets in terms of type-2 fuzzy logic systems, and (3) Use generalized.

Chapter 6 investigates type-2 fuzzy Markov random fields (T2 FMRFs) and their applications to handwritten Chinese character recognition. Stroke segmentation is formulated as the optimal labeling problem. We can use the MRF to detect the ambiguous parts. The MRF can describe the stroke relationships of Chinese characters statistically. The IT2 FMRF shows a better generalization ability than the classical MRF.

Chapter 7 proposes type-2 fuzzy topic models (T2 FTMs) and their applications to topic modeling and human action recognition. We propose a new algorithm, belief propagation (BP), for learning latent Dirichlet allocation (LDA). We also discuss how to speed up BP by residual BP and active BP techniques. We implement two T2 FTMs to perform human action recognition. Results confirm the effectiveness of type-2 fuzzy extensions for differentiating similar actions.

## 8.2 Future Works

In this book we study the probabilistic graphical models and type-2 fuzzy sets for pattern recognition. The graphical models have been deeply explored for pattern recognition in the past forty years. The question is how to design a suitable graphical model for the problem at hand. Basically, we have three issues in mind: (1) Determine the neighborhood systems (or topology); (2) Design clique potentials (or distance measure); (3) Develop learning and decoding algorithms. Future works may include applications to protein and gene structure recognition.

Mathematically, the three-dimensional structure of type-2 fuzzy membership function provides us a powerful tool for many complex problems. Furthermore, the type-2 fuzzy sets can be integrated with other methods, such as neural networks, support vector machines, genetic algorithms, to render many new algorithms for handling uncertainty. As far as pattern recognition is concerned, we are still in need of methods to make decisions based on the type-2 fuzzy sets. In the meantime, how to reduce the computational complexity of the general type-2 fuzzy sets operations is an open problem. Future works may integrate type-2 fuzzy sets with deep learning framework for pattern recognition.

# Errata to: Type-2 Fuzzy Graphical Models for Pattern Recognition

Jia Zeng and Zhi-Qiang Liu

**Errata to:**
**J. Zeng and Z.-Q. Liu** *Type-2 Fuzzy Graphical Models*
*for Pattern Recognition*, **Studies in Computational**
**Intelligence, DOI 10.1007/978-3-662-44690-4**

The volume number for 'Type-2 Fuzzy Graphical Models for Pattern Recognition'
has been changed from '666' to '591'.

---

The online version of the original book can be found under DOI 10.1007/978-3-662-44690-4

---

J. Zeng (✉)
School of Computer Science and Technology, Soochow University, Suzhou, China
e-mail: jiazeng@suda.edu.cn

Z.-Q. Liu
School of Creative Media, City University of Hong Kong, Hong Kong, China
e-mail: SMZLIU@cityu.edu.hk

© Tsinghua University Press, Beijing and Springer-Verlag Berlin Heidelberg 2015     E1
J. Zeng and Z.-Q. Liu, *Type-2 Fuzzy Graphical Models for Pattern Recognition*,
Studies in Computational Intelligence 591, DOI 10.1007/978-3-662-44690-4_9

Printed in the United States
By Bookmasters